Quantum Theory of Conducting Matter

Shigeji Fujita and Kei Ito

Quantum Theory of Conducting Matter

Newtonian Equations of Motion for a Bloch Electron

Shigeji Fujita
Department of Physics
University at Buffalo
The State University of New York
Buffalo, NY 14260
fujita@buffalo.edu

Kei Ito
Research Division
The National Center for University Entrance Examinations
2-19-23 Komaba, Meguro
Tokyo 153-8501
Japan
ito@rd.dnc.ac.jp

ISBN 978-1-4419-2547-3 e-ISBN 978-0-387-74103-1

Printed on acid-free paper.

9 8 7 6 5 4 3 2 1

springer.com

Preface

The measurements of the Hall coefficient R_H and the Seebeck coefficient (thermopower) S are known to give the sign of the carrier charge q. Sodium (Na) forms a body-centered cubic (BCC) lattice, where both R_H and S are negative, indicating that the carrier is the "electron." Silver (Ag) forms a face-centered cubic (FCC) lattice, where the Hall coefficient R_H is negative but the Seebeck coefficient S is positive. This complication arises from the Fermi surface of the metal. The "electrons" and the "holes" play important roles in conducting matter physics. The "electron" ("hole"), which by definition circulates counterclockwise (clockwise) around the magnetic field (flux) vector $\mathbf{B}$ cannot be discussed based on the prevailing equation of motion in the electron dynamics: $\hbar d\mathbf{k}/dt = q(\mathbf{E} + \mathbf{v} \times \mathbf{B})$, where $\mathbf{k} = k$-vector, $\mathbf{E}$ = electric field, and $\mathbf{v}$ = velocity. The energy-momentum relation is not incorporated in this equation.

In this book we shall derive Newtonian equations of motion with a symmetric mass tensor. We diagonalize this tensor by introducing the principal masses and the principal axes of the inverse-mass tensor associated with the Fermi surface. Using these equations, we demonstrate that the "electrons" ("holes") are generated, depending on the curvature sign of the Fermi surface. The complicated Fermi surface of Ag can generate "electrons" and "holes," and it is responsible for the observed negative Hall coefficient R_H and positive Seebeck coefficient S. When the Fermi surface is nonspherical, the conduction electron moves anisotropically with different effective masses (m_1, m_2, m_3). The magnetic oscillations in the susceptibility χ and the magnetoresistance arise from the oscillatory density of states assocated with the Landau states upon the application of a magnetic field. The most direct probe of the Fermi surface can be made by observing the cyclotron resonance. A magnetic field is applied to a pure sample at liquid helium temperatures. The sign of the charge carrier can be determined by using the circularly polarized lasers. The data are analyzed in terms of Shockley's formula or its simplified version. Most often, the intrinsic effective masses for a semiconductor or metal can be determined directly after simple analyses.

In the present volume, we mainly deal with the behaviors of the fermionic conduction electrons. In the companion volume, called book2 in the text, superconductivity and quantum Hall effect are treated. The charge carriers in the supercurrent are the Cooper pairs, each composed of a pair of electrons bound by the phonon exchange attraction. The statistics of a composite particle with respect to the center-of-mass motion follows Ehrenfest–Oppenheimer–Bethe's rule: a composite moves as a fermion (boson) if it contains an odd (even) number of elementary fermions. Accordingly the Cooper pair moves as a boson since the pair contains two electrons. The different statistics generate very different behavior. The quantum Hall effect arises from the supercurrent generated in a two-dimensional system subject to a magnetic field.

The text is composed of three parts: preliminaries, Bloch electron dynamics, and applications (fermionic systems). Part I, Chapters 1 through 6, starts with an introduction and then deals with the phonons (quanta of lattice vibrations), the free-electron model, the kinetic theory of electron transport, the magnetic susceptibility, and the Boltzmann equation method. These materials are normally covered in introductory solid-state physics courses; however, they are prerequisite to the theoretical developments in Part II, Chapters 7 through 10. The Bloch theorem, the self-consistent mean field theory, the Fermi surface, the Bloch electron (wave packet) dynamics with Newtonian equations of motion are discussed in Part II. In Part III, Chapters 11 through 15, a selection of applications for fermionic systems (mostly electrons) are discussed: the de Haas–van Alphen oscillations in susceptibility, the Shubnikov–de Haas oscillations in magnetoresistance, the angle-dependent cyclotron resonance, the Seebeck coefficient arising from the thermal diffusion, and the infrared-laser Faraday rotation.

The present book is written for first-year graduate students in physics, chemistry, electrical engineering, and material sciences. Dynamics, quantum mechanics, thermodynamics, statistical mechanics, electromagnetism, and solid-state physics at the undergraduate level are prerequisite. The authors believe that the students should learn physics, starting from the bottom up and following all theoretical developments with step-by-step calculations. We have included many problems, most of them elementary excercises in the text. The students learn key concepts more firmly by working out these problems.

The prevalent equations of motion for the electron in a crystal is challenged in this book, but all arguments leading to the new equations of motion are based on the principles of quantum statistical mechanics (Heisenberg uncertainty and Pauli exclusion principles). Condensed matter physicists, chemists, and material scientists, theoretical and experimental, are invited to examine this text.

The authors thank the following indivisuals for valuable criticisms, discussions, and readings: Professor M. de Llano, Universidad Nacional Autónoma de Mexico; Professor T. Obata, Gunma National College of Technology, Professor Robert Kohler, Buffalo State University; and Dr. Hung-Chuk Ho, Notre Dame University. They also thank Sachiko, Michio, Isao, Yoshiko, Eriko, George Redden, Karen Roth, and Kurt Borchardt for their encouragement and reading of the drafts.

Shigeji Fujita, Buffalo, New York, USA
and
Kei Ito, Tokyo, Japan
August 2007

Contents

III Applications. Fermionic Systems (Electrons) 131

Constants, Signs, Symbols, and General Remarks

Useful Physical Constants

Quantity	Symbol	Value
Absolute zero on Celsius scale		$-273.16°C$
Avogadro's number	N_0	6.02×10^{23} mol^{-1}
Boltzmann constant	k_B	1.38×10^{-16} erg K^{-1}
Bohr magneton	μ_B	9.22×10^{-21} erg gauss^{-1}
Bohr radius	a_0	5.29×10^{-9} cm
Electron mass	m	9.11×10^{-28} g
Electron charge (magnitude)	e	4.80×10^{-10} esu
Gas constant	R	8.314 J mol^{-1} K^{-1}
Molar volume (gas at STP)		2.24×10^{4} cm^3 = 22.4 liter
Mechanical equivalent of heat		4.186 J cal^{-1}
Permeability constant	μ_0	1.26×10^{-6} H m^{-1}
Permittivity constant	ϵ_0	8.854×10^{-12} F m^{-1}
Planck's constant	h	6.63×10^{-27} erg sec
Planck's constant/2π	$\hbar$	1.05×10^{-27} erg sec
Proton mass	m_p	1.67×10^{-24} g
Speed of light	c	3.00×10^{10} cm sec^{-1}

Mathematical Signs

Sign	Means
$=$	equals
$\simeq$	approximately equals
$\neq$	not equal to
$\equiv$	identical to, defined as
$>$	greater than
$\gg$	much greater than
$<$	less than
$\ll$	much less than
$\geq$	greater than or equal to
$\leq$	less than or equal to
$\propto$	proportional to
$\sim$	represented by, of the order
$\langle x \rangle$, $\bar{x}$	the average value of x
ln	natural logarithm
Δx	increment in x
dx	infinitesimal increment in x
$z^* = x - iy$	complex conjugate of $z = x + iy$, real (x, y)
$\alpha^\dagger$	Hermitian conjugate of operator (matrix) α
α^T	transpose of matrix α
P^{-1}	inverse of P
$\delta_{a,b} = \begin{cases} 1 & \text{if } a = b \\ 0 & \text{if } a \neq b \end{cases}$	Kronecker's delta
$\delta(x)$	Dirac's delta function
∇	nabla or del operator
$\dot{x} \equiv dx/dt$	time derivative
$\mathrm{grad}\,\phi \equiv \nabla\phi$	gradient of ϕ
$\mathrm{div}\mathbf{A} \equiv \nabla \cdot \mathbf{A}$	divergence of $\mathbf{A}$
$\mathrm{curl}\,\mathbf{A} \equiv \nabla \times \mathbf{A}$	curl of $\mathbf{A}$
∇^2	Laplacian operator

List of Symbols[1]

Symbol	Quantity
Å	Ångstrom (= 10^{-8} cm = 10^{-10} m)
$\mathbf{A}$	vector potential
$\mathbf{B}$	magnetic field (magnetic flux density)
C_V	heat capacity at constant volume
c	specific heat
c	speed of light
$\mathcal{D}(p)$	density of states in momentum space
$\mathcal{D}(\omega)$	density of states in angular frequency
E	total energy
E	internal energy
$\mathbf{E}$	electric field
e	base of natural logarithm
e	electronic charge (absolute value)
F	Helmholtz free energy
f	one-body distribution function
f_B	Bose distribution function
f_F	Fermi distribution function
f_0	Planck distribution function
G	Gibbs free energy
H	Hamiltonian
H_c	critical magnetic field (magnitude)
$\mathbf{H}_a$	applied magnetic field vector
$\mathcal{H}$	Hamiltonian density
h	Planck's constant
h	single-particle Hamiltonian

[1] The following list is not intended to be exhaustive. It includes symbols of special importance.

Symbol	Quantity
$\hbar$	Planck's constant divided by 2π
$i \equiv \sqrt{-1}$	imaginary number unit
$\mathbf{i}, \mathbf{j}, \mathbf{k}$	Cartesian unit vectors
J	Jacobian of transformation
$\mathbf{J}$	total current
$\mathbf{j}$	single-particle current
$\mathbf{j}$	current density
$\mathbf{k}$	angular wave vector $\equiv$ k-vector
k_B	Boltzmann constant
L	Lagrangian function
L	normalization length
ln	natural logarithm
$\mathcal{L}$	Lagrangian density
l	mean free pass
M	molecular mass
m	electron mass
m^*	effective mass
N	number of particles
$\hat{N}$	number operator
$\mathcal{N}(\epsilon)$	density of states in energy
n	particle number density
P	pressure
$\mathbf{P}$	total momentum
$\mathbf{p}$	momentum vector
p	momentum (magnitude)
Q	quantity of heat
R	resistance
$\mathbf{R}$	position of the center of mass
r	radial coordinate

Symbol	Quantity
$\mathbf{r}$	position vector
S	entropy
T	kinetic energy
T	absolute temperature
T_c	critical (condensation) temperature
T_F	Fermi temperature
t	time
TR	sum of N-particle trace $\equiv$ grand ensemble trace
Tr	many-particle trace
tr	one-particle trace
V	potential energy
V	volume
$\mathbf{v}$	velocity (field)
W	work
Z	partition function
$e^{\alpha} \equiv z$	fugacity
$\beta \equiv (k_B T)^{-1}$	reciprocal temperature
Δx	small variation in x
$\delta(x)$	Dirac delta-function
$\delta_P = \begin{cases} +1 & \text{if } P \text{ is even} \\ -1 & \text{if } P \text{ is odd} \end{cases}$	parity sign of the permutation P
ϵ	energy
ϵ_F	Fermi energy
η	viscosity coefficient
Θ_D	Debye temperature
Θ_E	Einstein temperature
θ	polar angle
λ	wavelength
λ	penetration depth

Symbol	Quantity
κ	curvature
μ	linear mass density of a string
μ	chemical potential
μ_B	Bohr magneton
ν	frequency = inverse of period
Ξ	grand partition function
ξ	dynamical variable
ξ	coherence length
ρ	mass density
ρ	density operator, system density operator
ρ	many-particle distribution function
σ	total cross section
σ	electrical conductivity
$\sigma_x, \sigma_y, \sigma_z$	Pauli spin matrices
τ	tension
τ_d	duration of collision
τ_c	average time between collisions
ϕ	azimuthal angle
ϕ	scalar potential
Ψ	quasi wave function for many condensed bosons
ψ	wave function for a quantum particle
$d\Omega = \sin\theta \, d\theta \, d\phi$	element of solid angle
$\omega \equiv 2\pi\nu$	angular frequency
ω_c	rate of collision
ω_D	Debye frequency
$[\,,\,]$	commutator brackets
$\{\,,\,\}$	anticommutator brackets
$\{\,,\,\}$	Poisson brackets
$[A]$	dimension of A

Compacted expression

If A is B (non-B), C is D (non-D) means that if A is B, C is D (if A is non-B, C is non-D).

Crystallographic notation

This is mainly used to denote a direction, or the orientation of a plane, in a cubic metal. A plane (hkl) intersects the orthogonal Cartesian axes, coinciding with the cube edges, at a/h, a/k, and a/l from the origin, a being a constant, usually the length of a side of the unit cell. The direction of a line is denoted by $[hkl]$, the direction cosines with respect to the Cartesian axes being h/N, k/N, and l/N, where $N^2 = h^2+k^2+l^2$. The indices may be separated by commas to avoid ambiguity. Only occasionally will the notation be used precisely; thus, [100] or [001] usually means any cube axis and [111], any diagonal.

B and H

When an electron is described in quantum mechanics, its interaction with a magnetic field is determined by **B** rather than **H**; that is, if the permeability μ is not unity, the electron motion is determined by $\mu\mathbf{H}$. It is preferable to forget **H** altogether and use **B** to define all field strengths. The vector potential **A** is correspondingly defined such that $\nabla \times \mathbf{A} = \mathbf{B}$. **B** is effectively the same inside and outside the metal sample.

Units

In much of the literature quoted, the unit of magnetic field **B** is the gauss. Electric fields are frequently expressed in V/cm and resistivities in Ω cm.

$$1 \text{ tesla (T)} = 10 \text{ kilogauss} \quad 1\ \Omega\ \text{m} = 10^2\ \Omega\ \text{cm}.$$

The Planck's constant over 2π, $\hbar \equiv h/2\pi$, is used in dealing with an electron. The original Planck's constant h is used in dealing with a photon.

Boxed equation

Equations of major importance are boxed.

Part I

Preliminaries

We review basic physical concepts, lattice vibrations (phonons), free-electron model, kinetic theory of electric conduction, magnetic susceptibility, and the Boltzmann equation method in Part I, Chapters 1 through 6. These topics serve as the preliminaries for the later development of the theory of conduction electrons.

Chapter 1

Introduction

We review the basic concepts in the theory of solids.

1.1 Crystal Lattices

When a material forms a solid, the atoms align themselves in a special periodic structure called a *crystal lattice*. For example, in rock salt (NaCl), Na^+ and Cl^- ions occupy the sites of a *simple cubic* (SC) lattice alternately as shown in Figure 1.1. In sodium (Na), Na^+ ions form a *body-centered cubic* (BCC) lattice as shown in Figure 1.2. In solid Ar, neutral atoms form a *face-centered cubic* (FCC) lattice as shown in Figure 1.3.

Solids can be classified into metals, insulators, semiconductors, etc., according to the electric transport properties. They can also be classified according to the modes of binding. The readers interested in these aspects are encouraged to refer to elementary solid-state textbooks, such as Kittel's [1]. An overwhelming fact is that almost all materials crystallize below a certain temperature. The exceptions are liquid He^3 and He^4, which do not freeze at atmospheric pressure. The thermal properties of crystalline solids are remarkably similar for all materials. At high temperatures *Dulong–Petit's law* for the heat capacity, $C_V = 3Nk_B$, is observed. The Boltzmann constant k_B has the following numerical value: $k_B = 1.38 \times 10^{-16}$ erg/degree $=$ 1.38×10^{-23} J K^{-1}. Deviations from this law become significant at low temperatures and are discussed further in Chapter 2.

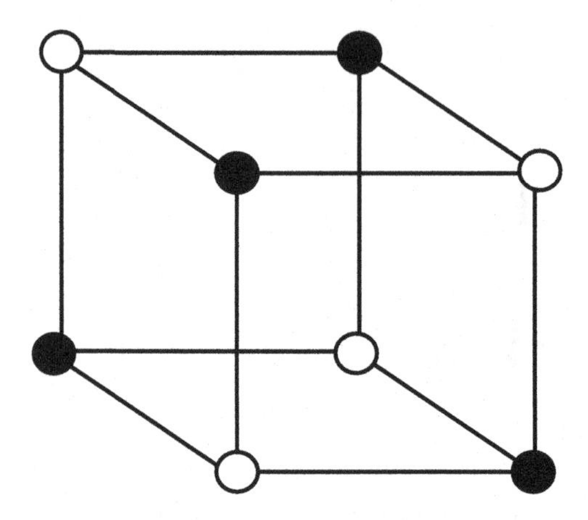

Figure 1.1: A simple cubic lattice. In rock salt, Na^+ (•) and Cl^- (○) occupy the lattice sites alternately.

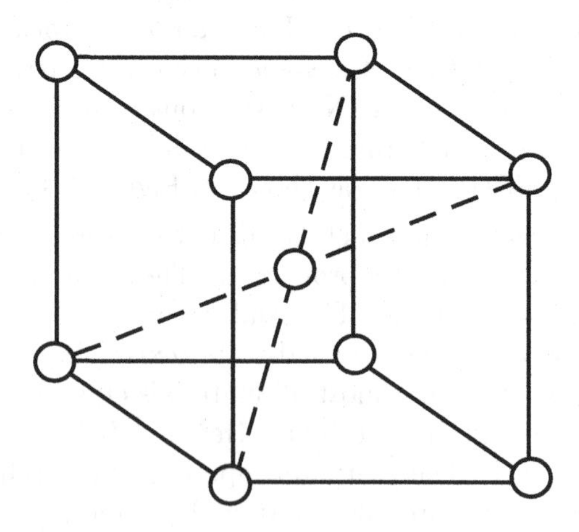

Figure 1.2: A body-centered cubic lattice.

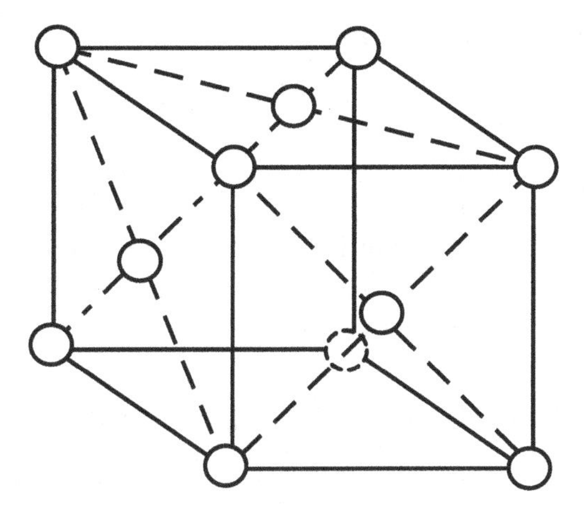

Figure 1.3: A face-centered cubic lattice.

1.2 Theoretical Background

1.2.1 Metals and Conduction Electrons

A metal is a conducting crystal in which electrical current can flow with little resistance. This electrical current is generated by moving electrons. The electron has mass m and charge $-e$, which is negative by convention. Their numerical values are $m = 9.1 \times 10^{-28}$ g and $e = 4.8 \times 10^{-10}$ esu $= 1.6 \times 10^{-19}$ C. The electron mass is about 1837 times smaller than the least massive (hydrogen) atom. This makes the electron extremely mobile. It also makes the electron's quantum nature more pronounced. The electrons participating in the charge transport are called *conduction electrons.* The conduction electrons would have orbited in the outermost shells surrounding the atomic nuclei if the nuclei were separated from each other. Core electrons which are more tightly bound with the nuclei form part of the metallic ions. In a pure crystalline metal, these metallic ions form a relatively immobile array of regular spacing, called a *lattice.* Thus, a metal can be pictured as a system of two components: mobile electrons and relatively immobile lattice ions.

1.2.2 Quantum Mechanics

Electrons move following the quantum laws of motion. A thorough understanding of quantum theory is essential. *Dirac's formulation* of quantum theory in his book, *Principles of Quantum Mechanics* [2], is unsurpassed. Dirac's rules that the quantum states are represented by *bra* or *ket* vectors and physical observables by Hermitian operators are used in the text. There are two distinct quantum effects, the first of which concerns a single particle and the second a system of identical particles.

1.2.3 Heisenberg Uncertainty Principle

Let us consider a simple harmonic oscillator characterized by the Hamiltonian

$$H = \frac{p^2}{2m} + \frac{kx^2}{2}, \tag{1.1}$$

where m is the mass, k the force constant, p the momentum, and x the position. The corresponding energy eigenvalues are

$$\epsilon_n = \hbar\omega_0 \left(n + \frac{1}{2}\right), \qquad \omega_0 \equiv \left(\frac{k}{m}\right)^{1/2}, \qquad n = 0, 1, 2, \ldots. \tag{1.2}$$

The energies are quantized in Equation (1.2). In contrast the classical energy can be any positive value. The lowest quantum energy $\epsilon_0 = \hbar\omega_0/2$, called the *energy of zero point motion*, is not zero. The most stable state of any quantum system is not a state of *static equilibrium* in the configuration of lowest potential energy, it is rather a *dynamic equilibrium* for the zero point motion [3]. Dynamic equilibrium may be characterized by the minimum total (potential + kinetic) energy under the condition that each coordinate q has a range Δq and the corresponding momentum p has a range Δp, so that the product $\Delta q \Delta p$ satisfies the *Heisenberg uncertainty relation*:

$$\Delta q \Delta p > h. \tag{1.3}$$

The most remarkable example of a macroscopic body in dynamic equilibrium is *liquid helium* (He). This liquid with a boiling point at 4.2 K is known to remain liquid down to 0 K. The zero-point motion of He atoms precludes solidification.

1.2.4 Bosons and Fermions

Electrons are fermions. That is, they are *indistinguishable quantum particles* subject to the *Pauli exclusion principle.* Indistinguishability of the particles is defined by using the permutation symmetry. According to Pauli's principle no two electrons can occupy the same state. Indistinguishable quantum particles not subject to the Pauli exclusion principle are called *bosons.* Bosons can occupy the same state with no restriction. *Every elementary particle is either a boson or a fermion.* This is known as the *quantum statistical postulate.* Whether an elementary particle is a boson or fermion is related to magnitude of its spin angular momentum in units of $\hbar$. *Particles with integer spins are bosons, while those with half-integer spins are fermions* [4]. This is known as Pauli's *spin-statistics theorem.* According to this theorem and in agreement with all experimental evidence, electrons, protons, neutrons, and μ-mesons, all of which have spin of magnitude $\hbar/2$, are fermions, while photons (quanta of electromagnetic radiation) with spin of magnitude $\hbar$, are bosons.

1.2.5 Fermi and Bose Distribution Functions

The average occupation number at state a, denoted by $\langle N_a \rangle$, for a system of free fermions in equilibrium at temperature T and chemical potential μ is given by the *Fermi distribution function*:

$$\langle N_a \rangle = f_F(\epsilon_a) \equiv \frac{1}{\exp[(\epsilon_a - \mu)/k_B T] + 1} \quad \text{for fermions,} \tag{1.4}$$

where ϵ_a is the energy associated with the state a.

The average occupation number at state for a system of free bosons in equilibrium is given by the *Bose distribution function*:

$$\langle N_a \rangle = f_B(\epsilon_a) \equiv \frac{1}{\exp[(\epsilon_a - \mu)/k_B T] - 1} \quad \text{for bosons.} \tag{1.5}$$

1.2.6 Composite Particles

Atomic nuclei are composed of *nucleons* (protons, neutrons), while atoms are composed of nuclei and electrons. It has been experimentally demonstrated that these composite particles are indistinguishable quantum particles. According to Ehrenfest–Oppenheimer–Bethe's rule [5] the center of mass (CM) of a composite moves as a fermion (boson) if it contains an odd (even) number of elementary fermions. Thus He^4 atoms (4 nucleons, 2 electrons) move as bosons and He^3 atoms (3 nucleons, 2 electrons) as fermions. Cooper pairs (2 electrons) move as bosons.

1.2.7 Quasifree Electron Model

In a metal at the lowest temperatures conduction electrons move in a nearly stationary periodic lattice. Because of the Coulomb interaction among the electrons, the motion of the electrons is correlated. However each electron in a crystal moves in an extremely weak self-consistent periodic field. Combining this result with the Pauli exclusion principle, which applies to electrons with no regard to the interaction, we obtain the *quasifree electron model*. The quasifree electron moves with the effective mass m^* different from the gravitational mass m_e. The subscript e is often omitted in the text: $m_e = m$. In this model the quantum states for the electron in a crystal are characterized by k-vector $\mathbf{k}$ and energy

$$\epsilon = E(k). \tag{1.6}$$

At 0 K, all of the lowest energy states are filled with electrons, and there exists a *sharp Fermi surface* represented by

$$E(k) = \epsilon_F, \tag{1.7}$$

where ϵ_F is the *Fermi energy*. Experimentally, the electrons in alkali (BCC) metals including lithium (Li), sodium (Na), and potassium (K) behave like quasifree electrons.

1.2.8 "Electrons" and "Holes"

"Electrons" (*"holes"*) in the text are defined as quasiparticles possessing charge e (magnitude) that circulate counterclockwise (clockwise) when viewed from the tip of the applied magnetic field vector $\mathbf{B}$. This definition is used routinely in semiconductor physics. We use quotation-marked "electron" to distinguish it from the generic electron having the gravitational mass m_e. A "hole" can be regarded as a particle having positive charge, positive mass, and positive energy. The "hole" does not, however, have the same effective

mass m^* (magnitude) as the "electron," so that "holes" are not true antiparticles like positrons. We will see that "electrons" and "holes" are the thermally excited particles and they are closely related to the curvature of the Fermi surface (see Sections 4.4 and 10.1).

Chapter 2

Lattice Vibrations and Heat Capacity

Einstein's and Debye's theories of the heat capacity of a solid are presented. The quantum particles (phonons) for the lattice vibrations are introduced.

2.1 Einstein's Theory of Heat Capacity

Let us consider a crystal lattice at low temperatures. We may expect each atom forming the lattice to execute small oscillations around the equilibrium position. For illustration let us consider the one-dimensional lattice shown in Figure 2.1. The motion of the jth atom may be characterized by the Hamiltonian of the form:

$$h_j = \frac{p_j^2}{2M} + \frac{1}{2}k_0 u_j^2, \tag{2.1}$$

where u_j denotes displacement and $p_j = M\dot{u}_j$ represents the momentum. Here we assumed a parabolic potential, which is reasonable for small oscillations. If the *equipartition theorem* is applied, the kinetic and potential energy parts each contribute $k_B T/2$ to the average thermal energy. We then obtain

$$\langle h \rangle = \left\langle \frac{1}{2M}p^2 \right\rangle + \left\langle \frac{1}{2}k_0 u^2 \right\rangle = \frac{1}{2}k_B T + \frac{1}{2}k_B T = k_B T. \tag{2.2}$$

Multiplying Equation (2.2) by the number of atoms N, we obtain Nk_BT for the total energy. By differentiating it with respect to T, we obtain the heat capacity Nk_B for the lattice.

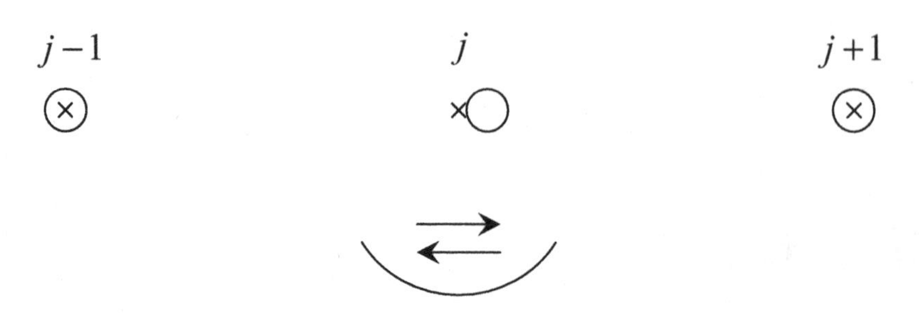

Figure 2.1: The jth atom in the linear chain executes simple harmonic oscillations as characterized by the Hamiltonian in Equation (2.1).

In quantum theory the eigenvalues of the Hamiltonian

$$h = \frac{p^2}{2M} + \frac{1}{2}k_0 u^2 \tag{2.3}$$

are given by

$$\epsilon_n = \left(\frac{1}{2} + n\right)\hbar\omega_0, \qquad n = 0, 1, 2, \ldots, \tag{2.4}$$

$$\omega_0 \equiv \left(\frac{k_0}{M}\right)^{1/2}. \tag{2.5}$$

The quantum states are characterized by nonnegative integers n. If we assume a canonical ensemble of simple harmonic oscillators with the distribution $\exp(-\beta\epsilon)$, the average energy can be calculated as follows:

$$\begin{aligned}\langle h \rangle &= \frac{\sum_n \epsilon_n e^{-\beta\epsilon_n}}{\sum_n e^{-\beta\epsilon_n}} = -\frac{\partial}{\partial\beta}\ln\left(\sum_n e^{-\beta\epsilon_n}\right) = -\frac{\partial}{\partial\beta}\ln\left(\sum_{n=0}^{\infty} e^{-\beta(1/2+n)\hbar\omega_0}\right) \\ &= -\frac{\partial}{\partial\beta}\ln\left(\frac{e^{-\beta\hbar\omega_0/2}}{1 - e^{-\beta\hbar\omega_0}}\right) = \left[\frac{1}{2} + f_0(\hbar\omega_0)\right]\hbar\omega_0,\end{aligned} \tag{2.6}$$

$$f_0(\eta) \equiv \frac{1}{e^{\beta\eta} - 1}. \tag{2.7}$$

Here f_0 represents the *Planck distribution function*. Notice that the quantum average energy $\langle h \rangle$ in Equation (2.6) is quite different from the classical average energy $k_B T$ [see Equation (2.2)].

Let us now consider N atoms in a three-dimensional lattice. We multiply Equation (2.6) by the number of degrees of freedom, $3N$ and obtain for the total energy

$$E = 3N \left[\frac{1}{2} + f_0(\hbar\omega_0) \right] \hbar\omega_0. \tag{2.8}$$

Differentiating this with respect to T, we obtain the heat capacity at constant volume C_V (Problem 2.1.1):

$$C_V = \frac{\partial E}{\partial T} = \frac{\partial E}{\partial \beta}\frac{\partial \beta}{\partial T} = \frac{3N(\hbar\omega_0)^2}{k_B T^2} \frac{e^{\beta\hbar\omega_0}}{\left(e^{\beta\hbar\omega_0} - 1\right)^2}.$$

We rewrite this expression in the form (Problem 2.1.1)

$$\boxed{C_V = 3Nk_B \left(\frac{\Theta_E}{T} \right)^2 e^{\Theta_E/T} \left(e^{\Theta_E/T} - 1 \right)^{-2},} \tag{2.9}$$

where

$$\Theta_E \equiv \frac{\hbar\omega_0}{k_B} = \left(\frac{\hbar}{k_B} \right) \left(\frac{k_0}{M} \right)^{1/2} \tag{2.10}$$

is the *Einstein temperature.* The function

$$g(x) \equiv x^2 e^x \left(e^x - 1 \right)^{-2} \qquad \left(x \equiv \frac{\Theta_E}{T} \right) \tag{2.11}$$

has the asymptotic behavior (Problem 2.1.2)

$$g(x) \approx \begin{cases} x^2 e^{-x} & \text{for } x \gg 1 \\ 1 & \text{for } x \ll 1. \end{cases} \tag{2.12}$$

At very high temperatures ($\Theta_E/T \ll 1$), the heat capacity given by Equation (2.9) approaches the classical value $3Nk_B$:

$$C_V = 3Nk_B g\left(\frac{\Theta_E}{T} \right) \longrightarrow 3Nk_B. \tag{2.13}$$

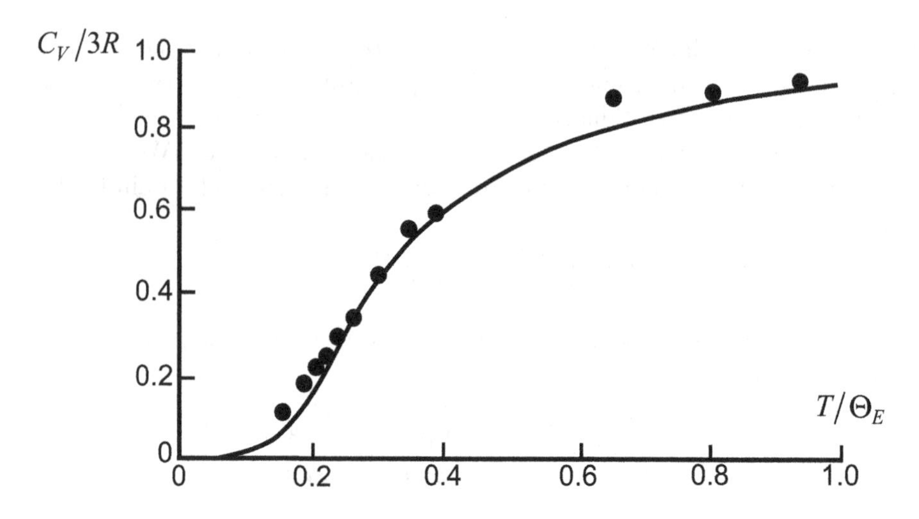

Figure 2.2: The molar heat capacity for Einstein's model of a solid is shown in a solid line. Experimental points are for a diamond with $\Theta_E = 1320$ K.

At very low temperatures ($\Theta_E/T \gg 1$), the heat capacity C_V behaves like

$$C_V \approx 3Nk_B \left(\frac{\Theta_E}{T}\right)^2 \exp\left(\frac{-\Theta_E}{T}\right) \tag{2.14}$$

and approaches zero exponentially as T tends to zero. The behavior of $C_V/3Nk_B$ from Equation (2.9) is plotted against Θ_E/T in Figure 2.2. Experimental data of C_V for diamond are shown by points with the choice of $\Theta_E = 1320$ K for comparison. The fit is quite reasonable for diamond. Historically Einstein [1] obtained Equation (2.9) in 1907 before the advent of quantum theory in 1925–26. The discrepancy between experiment and theory (equipartition theorem) had been a mystery until 1907.

Problem 2.1.1. Derive Equation (2.9), using Equations (2.7) and (2.8).

Problem 2.1.2. Verify the asymptotic behavior of Equation (2.12) from Equation (2.11).

2.2 Debye's Theory of Heat Capacity

In Section 2.1, we discussed Einstein's theory of the heat capacity of a solid. This theory explains why a certain solid like a diamond exhibits a molar heat capacity substantially smaller than $3R$. It also predicts that the heat capacity for any solid should decrease as the temperature is lowered. This prediction is in agreement with experimental observations. However, the temperature dependence at very low temperatures indicates nonnegligible discrepancies. In 1912 Debye [2] reported a very successful theory of the heat capacity of solids, which will be discussed in this section.

In Einstein's model the solid is viewed as a collection of free harmonic oscillators with a *common* frequency ω_0. This is rather a drastic approximation, because an atom in a real crystal moves under the varying potential field generated by its neighboring atoms. It is more reasonable to look at the crystal as a collection of *coupled* harmonic oscillators. The motion of a set of coupled harmonic oscillators can be described concisely in terms of the normal modes of oscillations. A major difficulty of this approach, however, is to find the set of the normal-mode frequencies. Debye overcame this difficulty by regarding a solid as a *continuous elastic body* whose normal modes of oscillations are known as *elastic waves.* We note that this is valid only at low frequencies.

A macroscopic solid can support transverse and longitudinal waves. We first look at the case of the transverse wave. The equation of motion for the transverse elastic wave is

$$\rho \frac{\partial^2 \mathbf{u}}{\partial t^2} = S \nabla^2 \mathbf{u}, \tag{2.15}$$

where $\mathbf{u}$ is the transverse displacement lying in a plane perpendicular to the direction of the wave propagation, ρ is the mass density, and S the shear modulus. Since the displacement (component) u_j is a continuous-field (opposed to a discrete) variable, the usual Lagrangian formulation does not work. However its generalization is immediate. We use a *Lagrangian field method* as follows.

We introduce a Lagrangian density $\mathcal{L}$, which is a function of the displacement components, their time-derivatives, and space-derivatives:

$$\mathcal{L} = \mathcal{L}\left(u_x, u_y, u_z, \frac{\partial u_x}{\partial t}, \frac{\partial u_y}{\partial t}, \frac{\partial u_z}{\partial t}, \frac{\partial u_x}{\partial x}, \dots\right). \tag{2.16}$$

We then derive the equations of motion by means of *Lagrange's field equations*:

$$\frac{\partial}{\partial t}\left(\frac{\partial \mathcal{L}}{\partial\left(\partial u_a/\partial t\right)}\right)+\frac{\partial}{\partial x}\left(\frac{\partial \mathcal{L}}{\partial\left(\partial u_a/\partial x\right)}\right)+\frac{\partial}{\partial y}\left(\frac{\partial \mathcal{L}}{\partial\left(\partial u_a/\partial y\right)}\right)$$
$$+\frac{\partial}{\partial z}\left(\frac{\partial \mathcal{L}}{\partial\left(\partial u_a/\partial z\right)}\right)-\frac{\partial \mathcal{L}}{\partial u_a}=0, \quad a=x,y,z. \tag{2.17}$$

In our case we may choose

$$\mathcal{L} = \frac{1}{2}\rho\left|\frac{\partial \mathbf{u}}{\partial t}\right|^2-\frac{1}{2}S\left(\left|\frac{\partial \mathbf{u}}{\partial x}\right|^2+\left|\frac{\partial \mathbf{u}}{\partial y}\right|^2+\left|\frac{\partial \mathbf{u}}{\partial z}\right|^2\right) \tag{2.18}$$

$$= \frac{1}{2}\rho\left[\left(\frac{\partial u_x}{\partial t}\right)^2+\left(\frac{\partial u_y}{\partial t}\right)^2+\left(\frac{\partial u_z}{\partial t}\right)^2\right]-\cdots, \tag{2.19}$$

and verify the correct equation of motion (2.15) by using Equations (2.17) and (2.19) (Problem 2.2.1).

Let us assume a cubic box (with the side length L_0) fixed-end boundary condition. The normal modes of oscillations are represented by the amplitude functions

$$\sin(xk_x)\sin(yk_y)\sin(zk_z), \tag{2.20}$$

$$k_a \equiv \frac{\pi n_a}{L_0}, \quad n_a=1,2,\ldots, \quad a=x,y,z, \tag{2.21}$$

and the angular frequencies

$$\omega_{t,k}=c_t\left|\mathbf{k}\right|=\left(\frac{S}{\rho}\right)^{1/2}k, \tag{2.22}$$

where c_t is the speed of propagation for the transverse (t) wave. The wave vector $\mathbf{k}$ is perpendicular to the displacement $\mathbf{u}$:

$$\mathbf{k}\cdot\mathbf{u}=0 \qquad \text{(transverse wave)}. \tag{2.23}$$

We now wish to express the Lagrangian density $\mathcal{L}$ in terms of the normal coordinates and velocities. Let us denote the Cartesian components of $\mathbf{u}$ by u_1 and u_2. We introduce the *normal coordinates*:

$$q_{\mathbf{k},\sigma}(t)=\left(\frac{2}{L_0}\right)^3\int_0^{L_0}dx\int_0^{L_0}dy\int_0^{L_0}dz\,u_\sigma\sin(xk_x)\sin(yk_y)\sin(zk_z), \tag{2.24}$$

where $\sigma = 1, 2$ are called the *polarization indices.* After lengthy but straightforward calculations, we can then transform the Lagrangian density $\mathcal{L}$ into (Problem 2.2.2)

$$\mathcal{L} = \sum_{\mathbf{k}} \sum_{\sigma} \left(\frac{1}{4} \rho \dot{q}_{\mathbf{k},\sigma}^2 - \frac{1}{4} S k^2 q_{\mathbf{k},\sigma}^2 \right), \tag{2.25}$$

which is the sum of single-mode Lagrangian densities: $(1/4)\rho \dot{q}_{\mathbf{k},\sigma}^2 - (1/4) S k^2 q_{\mathbf{k},\sigma}^2$.

The Lagrangian L is the Lagrangian density $\mathcal{L}$ times the volume V:

$$L \equiv V \mathcal{L}. \tag{2.26}$$

For later convenience we introduce new normal coordinates:

$$Q_{\mathbf{k},\sigma} \equiv \left(\frac{V\rho}{2} \right)^{1/2} q_{\mathbf{k},\sigma}. \tag{2.27}$$

By using these coordinates, we can re-express the Lagrangian L as follows:

$$L = \sum_{\mathbf{k}} \sum_{\sigma} \frac{1}{2} \left[\dot{Q}_{\mathbf{k},\sigma}^2 - \omega_{t,k}^2 Q_{\mathbf{k},\sigma}^2 \right]. \tag{2.28}$$

To derive the corresponding Hamiltonian, we define the *canonical momenta* $\{P_{\mathbf{k},\sigma}\}$ by

$$P_{\mathbf{k},\sigma} \equiv \frac{\partial L}{\partial \dot{Q}_{\mathbf{k},\sigma}} = \dot{Q}_{\mathbf{k},\sigma}. \tag{2.29}$$

The Hamiltonian H_t for the transverse waves can then be constructed by expressing $\sum_{\mathbf{k}} \sum_{\sigma} P_{\mathbf{k},\sigma} \dot{Q}_{\mathbf{k},\sigma} - L$ in terms of $\{Q_{\mathbf{k},\sigma}, P_{\mathbf{k},\sigma}\}$. The result is given by

$$H_t = \sum_{\mathbf{k}} \sum_{\sigma} \frac{1}{2} \left[P_{\mathbf{k},\sigma}^2 + \omega_{t,k}^2 Q_{\mathbf{k},\sigma}^2 \right]. \tag{2.30}$$

The normal modes of oscillations characterized by k-vector $\mathbf{k}$, polarization σ, and frequency $\omega_{t,k}$ represent *standing waves*, which arise from the fixed-end boundary condition. If we assume a periodic box boundary condition, the normal modes are the *running waves*. In either case the results for the heat capacity are the same.

The energy eigenvalues for the corresponding quantum Hamiltonian for the system are

$$E_t = \sum_{\mathbf{k}} \sum_{\sigma} \left(\frac{1}{2} + n_{\mathbf{k},\sigma} \right) \hbar\omega_{t,k} = E\left[\{n_{\mathbf{k},\sigma}\}\right], \tag{2.31}$$

$$n_{\mathbf{k},\sigma} = 0, 1, 2, \ldots. \tag{2.32}$$

It is convenient to interpret Equation (2.31) in terms of the energies of quantum particles called *phonons*. In this view the quantum number $n_{\mathbf{k},\sigma}$ represents the number of phonons in the normal mode $(\mathbf{k}, \sigma)$. The energy of the system is then specified by the set of the numbers of phonons in normal modes (states), $\{n_{\mathbf{k},\sigma}\}$.

By taking the canonical ensemble average of Equation (2.31), we obtain

$$\begin{aligned} \langle E_t \rangle &= \sum_{\mathbf{k}} \sum_{\sigma} \left(\frac{1}{2} + \langle n_{\mathbf{k},\sigma} \rangle \right) \hbar\omega_{t,k} \\ &= \sum_{\mathbf{k}} \sum_{\sigma} \left[\frac{1}{2} + f_0\left(\hbar\omega_{t,k}\right) \right] \hbar\omega_{t,k} \equiv E_t(\beta), \end{aligned} \tag{2.33}$$

$$\boxed{\langle n_{\mathbf{k},\sigma} \rangle = \frac{1}{e^{\beta\epsilon} - 1} \equiv f_0(\epsilon).} \tag{2.34}$$

When the volume of normalization V is made large, the distribution of the normal-mode points in the k-space becomes dense. In the bulk limit we may convert the sum over $\{\mathbf{k}\}$ in Equation (2.33) into an ω-integral and obtain

$$E_t(\beta) = E_{t,0} + \int_0^\infty d\omega\, \hbar\omega\, f_0(\hbar\omega) \mathcal{D}_t(\omega), \tag{2.35}$$

$$E_{t,0} \equiv \frac{1}{2} \int_0^\infty d\omega\, \hbar\omega \mathcal{D}_t(\omega), \tag{2.36}$$

where $\mathcal{D}_t(\omega)$ is the *density of states (modes) in the frequency domain* defined such that

$$\text{Number of modes in the interval } (\omega, \omega + d\omega) \equiv \mathcal{D}_t\, d\omega, \tag{2.37}$$

and $E_{t,0}$ represents the sum of zero-point energies, which is independent of temperature.

The density of states in the frequency domain $\mathcal{D}_t(\omega)$ may be calculated as follows. First we note that the angular frequency ω is related to the wave vector $\mathbf{k}$ by $\omega = c_t k$. The constant-ω surface in the k-space is therefore the sphere of radius $k = \omega/c_t$. Consider another concentric sphere of radius $k+dk = (\omega+d\omega)/c_t$. Each mode point is located at the SC lattice sites with unit spacing π/L_0 and only in the first octant ($k_x, k_y, k_z > 0$). The number of mode points within the spherical shell between the two spheres can be obtained by dividing one-eighth of the k-volume of the shell:

$$\frac{(4\pi k^2\, dk)}{8} = \left(\frac{1}{2}\right)\frac{\pi\omega^2\, d\omega}{c_t^3} \qquad (\omega = c_t k)$$

by the unit-cell k-volume, $(\pi/L_0)^3$. Multiplying the result by 2 in consideration of the polarization multiplicity, we obtain

$$\mathcal{D}_t(\omega)\, d\omega = \frac{\pi(\omega^2\, d\omega/c_t^3)}{(\pi/L_0)^3} = \frac{L_0^3}{\pi^2}\frac{\omega^2\, d\omega}{c_t^3} = \frac{V}{\pi^2}\frac{\omega^2\, d\omega}{c_t^3}. \tag{2.38}$$

We may treat the longitudinal waves similarly. The average energy $E_l(\beta) \equiv \langle E_l\,[\{n_{\mathbf{k}}\}]\rangle$ can be written in the same form as Equation (2.33) with the density of states

$$\mathcal{D}_l(\omega) = \frac{V}{2\pi^2}\frac{\omega^2}{c_l^3}; \tag{2.39}$$

$$c_l \equiv \left[\frac{(B + 3S/4)}{\rho}\right]^{1/2} \tag{2.40}$$

is the propagation speed of the longitudinal wave; $B \equiv -V^{-1}\,\partial P/\partial V)_T$ is the bulk modulus. Combining the two wave modes together, we obtain

$$E \equiv E_t + E_l = E_0 + \int_0^\infty d\omega\, \hbar\omega\, f_0(\hbar\omega)\mathcal{D}(\omega), \tag{2.41}$$

$$E_0 \equiv \frac{1}{2}\int_0^\infty d\omega\, \hbar\omega\mathcal{D}(\omega), \tag{2.42}$$

where

$$\mathcal{D}(\omega) \equiv \mathcal{D}_t(\omega) + \mathcal{D}_l(\omega) = \frac{V}{2\pi^2}\left(\frac{2}{c_t^3} + \frac{1}{c_l^3}\right)\omega^2 \tag{2.43}$$

denotes the density of states in frequency for the combined wave modes.

For a real crystal of N atoms, the number of degrees of freedom is $3N$. Therefore, there exist exactly $3N$ normal modes. The elastic body whose dynamic state is described by the vector field $\mathbf{u}(\mathbf{r}, t)$ has an infinite number of degrees of freedom, and therefore, it has infinitely many modes. The model of a macroscopic elastic body is a reasonable representation of a real solid for low-frequency modes, that is, long-wavelength modes. At high frequencies the continuum model does not provide normal modes expected of a real solid. Debye solved the problem by assuming that the density of states $\mathcal{D}_D(\omega)$ has the same value as given by Equation (2.43) up to the maximum frequency ω_D, then vanishes thereafter,

$$\mathcal{D}_D(\omega) = \begin{cases} \dfrac{V}{2\pi^2}\left(\dfrac{2}{c_t^3} + \dfrac{1}{c_l^3}\right)\omega^2 & \omega < \omega_D \\ 0 & \text{otherwise,} \end{cases} \tag{2.44}$$

and that the limit frequency ω_D, called the *Debye frequency*, is chosen such that the number of modes for the truncated continuum model equals $3N$:

$$\int_0^\infty d\omega\, \mathcal{D}_D(\omega) = \int_0^{\omega_D} d\omega\, \mathcal{D}(\omega) = 3N. \tag{2.45}$$

Substitution of Equation (2.44) yields (Problem 2.2.3.)

$$\omega_D = \left[18\pi^2 n \Big/ \left(\frac{2}{c_t^3} + \frac{1}{c_l^3}\right)\right]^{1/3}, \tag{2.46}$$

where $n \equiv N/V$ is the number density. The density of states $\mathcal{D}_D(\omega)$ grows quadratically, reaches the maximum at $\omega = \omega_D$, then vanishes thereafter as shown in Figure 2.3.

We now compute the heat capacity C_V. Introducing Equation (2.44) into Equation (2.41), we obtain

$$\begin{aligned} E(T) - E_0 &= \int_0^{\omega_D} d\omega\, \hbar\omega f_0(\hbar\omega) \frac{V}{2\pi^2}\left(\frac{2}{c_t^3} + \frac{1}{c_l^3}\right)\omega^2 \\ &= \frac{9N\hbar}{\omega_D^3}\int_0^{\omega_D} d\omega \frac{\omega^3}{\exp(\hbar\omega/k_B T) - 1}. \end{aligned} \tag{2.47}$$

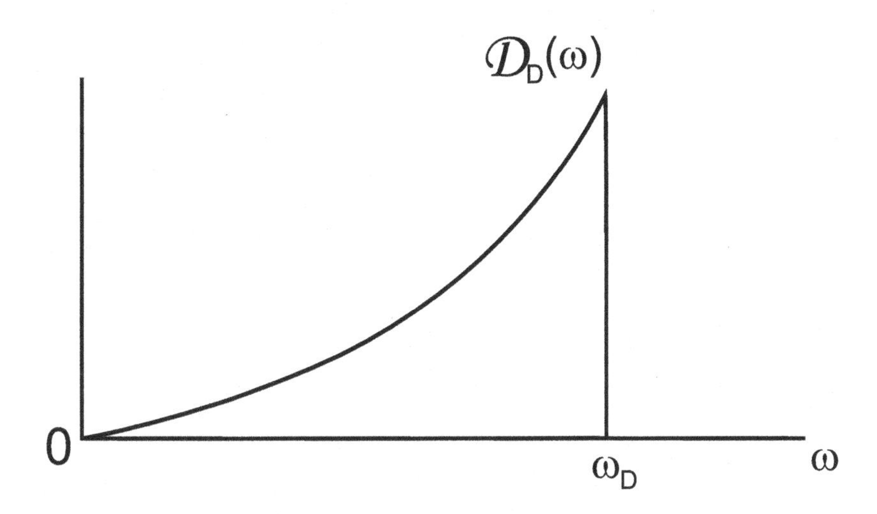

Figure 2.3: The density of modes in angular frequency for the Debye model.

Introducing the *Debye temperature* Θ_D defined by

$$\Theta_D \equiv \frac{\hbar\omega_D}{k_B}, \tag{2.48}$$

we can rewrite Equation (2.47) as

$$E(T) - E_0 = 9Nk_BT\left(\frac{T}{\Theta_D}\right)^3 \int_0^{x_D} dx \frac{x^3}{e^x - 1}, \tag{2.49}$$

$$x_D \equiv \frac{\hbar\omega_D}{k_BT} \equiv \frac{\Theta_D}{T}, \qquad x \equiv \frac{\hbar\omega}{k_BT}. \tag{2.50}$$

Consider first the high-temperature region where $x \equiv \hbar\omega/k_BT \ll 1$. Then $e^x - 1 \approx x$ and

$$\int_0^{x_D} dx \frac{x^3}{e^x - 1} \simeq \int_0^{x_D} dx\, x^2 = \frac{x_D^3}{3} = \frac{1}{3}\left(\frac{\Theta_D}{T}\right)^3, \qquad x_D \ll 1. \tag{2.51}$$

Using this approximation, we obtain from Equation (2.49)

$$E(T) = E_0 + 3Nk_BT, \qquad T \gg \Theta_D. \tag{2.52}$$

Differentiating Equation (2.52) with respect to T, we obtain the heat capacity of the solid, $C_V = \partial E/\partial T = 3Nk_B$, which is in agreement with the Dulong–Petit's law.

At very low temperatures, $x_D \equiv \hbar\omega_D/k_BT$ becomes very large. In the limit, we have

$$\begin{aligned} \int_0^{\omega_D} dx \frac{x^3}{e^x - 1} &\longrightarrow \int_0^{\infty} dx \frac{x^3}{e^x - 1} \\ &= \int_0^{\infty} dx\, x^3 e^{-x}(1 + e^{-x} + e^{-2x} + \cdots) \\ &= \sum_{n=1}^{\infty} \frac{6}{n^4} = \frac{\pi^4}{15}. \end{aligned} \tag{2.53}$$

We then obtain from Equation (2.49)

$$E = E_0 + \left(\frac{3}{5}\right) \frac{\pi^4 Nk_BT^4}{\Theta_D^3} \qquad (T \ll \Theta_D). \tag{2.54}$$

Differentiating this expression with respect to T, we obtain

$$\boxed{C_V = \frac{12\pi^2}{5} Nk_B \left(\frac{T}{\Theta_D}\right)^3.} \tag{2.55}$$

According to this equation, the heat capacity at very low temperatures behaves like T^3. This is known as the *Debye T^3 law*. The decrease is therefore slower than the Einstein heat capacity given by Equation (2.14), which decreases exponentially.

At intermediate temperatures the integral may be evaluated as follows. Differentiating Equation (2.47) with respect to T, we obtain

$$\boxed{C_V = 9Nk_B \left(\frac{T}{\Theta_D}\right)^3 \int_0^{x_D} dx \frac{x^4 e^x}{(e^x - 1)^2}} \quad \left(x_D \equiv \frac{\Theta_D}{T}\right). \tag{2.56}$$

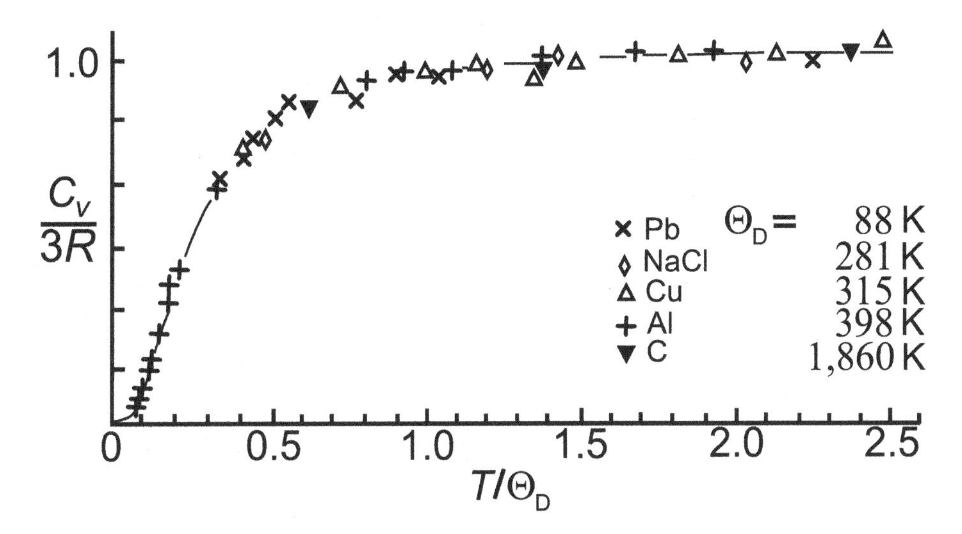

Figure 2.4: The heat capacities of solids. The solid line represents the Debye curve obtained from Equation (2.55). Experimental data for various solids are shown with optimum Debye temperatures, after Sears and Salinger [3].

We can compute this integral numerically. We then express the molar heat of solid as a function of T/Θ_D only. The theoretical curve obtained is shown in a solid line in Figure 2.4. The points in the figure are experimental data for various solids. The Debye temperatures adjusted at optimum fitting are also shown in the same figure. Roughly speaking when T/Θ_D is greater than 1 or when the actual temperature exceeds the Debye temperature, the solid behaves classically, and the molar heat is nearly equal to $3R$ in agreement with the Dulong–Petit's law. When the temperature is less than the Debye temperature, the quantum effect sets in; so the heat capacity is less than $3R$. Thus, for lead (a soft metal) with a Debye temperature of only 88 K, room temperature is well above the Debye temperature. Therefore, its heat capacity has a classical value. Diamond (a hard crystal) with a Debye temperature of 1860 K, shows a quantum nature in heat capacity even at room temperature. The Debye temperatures for several typical solids are given in Table 2.1.

Debye's theory of the heat capacity gives quite satisfactory results for almost all nonconducting crystals. Exceptions arise if the crystals have striking

Table 2.1: Approximate Debye Temperature Values

Solids	Debye temperature Θ_D in K
NaCl	281
KCl	230
Pb	88
Ag	225
Zn	308
Diamond	1860

anisotropies like graphite or if finite-size crystals are considered. For conducting materials we must take into account the electronic contribution to the heat capacity, which will be discussed in Chapter 3. We may look at a solid as a collection of coupled harmonic oscillators and develop a theory from this viewpoint.

Problem 2.2.1. Verify Equation (2.15) by using Equations (2.17) and (2.19).

Problem 2.2.2. Verify Equation (2.25).

Problem 2.2.3. Verify Equation (2.46).

Chapter 3

Free Electrons and Heat Capacity

The quantum theory of the free-electron model resolves the heat capacity paradox, that is, the apparent absence of electron contribution to heat capacity.

3.1 Free Electrons and the Fermi Energy

Let us consider a system of *free electrons* characterized by the Hamiltonian

$$H = \sum_j \frac{p_j^2}{2m}. \tag{3.1}$$

The momentum eigenstates for a quantum particle with a periodic cube-box boundary condition are characterized by three quantum numbers:

$$p_{x,j} = \left(\frac{2\pi\hbar}{L}\right) j, \quad p_{y,k} = \left(\frac{2\pi\hbar}{L}\right) k, \quad p_{z,l} = \left(\frac{2\pi\hbar}{L}\right) l, \tag{3.2}$$

where L is the cube side length and j, k, and l are integers. For simplicity, we indicate the momentum state by a single Greek letter κ:

$$\mathbf{p}_\kappa \equiv (p_{x,j},\, p_{y,k},\, p_{z,l}). \tag{3.3}$$

The quantum state of our many-electron system can be specified by the set of occupation numbers $\{n_\kappa\}$, with each $n_\kappa \equiv n_{\mathbf{p}_\kappa}$ taking on either one or zero. The ket vector representing such a state will be denoted by

$$|\{n\}\rangle \equiv |\{n_\kappa\}\rangle \,. \tag{3.4}$$

The corresponding energy eigenvalue is given by

$$E(\{n\}) = \sum_\kappa \epsilon_\kappa n_\kappa, \tag{3.5}$$

where $\epsilon_\kappa \equiv p_\kappa^2/2m$ is the kinetic energy of the electron with momentum $\mathbf{p}_\kappa$. The sum of the occupation numbers n_κ equals the total number N of electrons:

$$\sum_\kappa n_\kappa = N. \tag{3.6}$$

We assume that the system is in thermodynamic equilibrium, which is characterized by temperature $T \equiv (k_B\beta)^{-1}$ and number density n. The thermodynamic properties of the system can then be computed in terms of the *grand canonical density operator*:

$$\rho_G = \frac{e^{\alpha N - \beta H}}{\mathrm{TR}\left\{e^{\alpha N - \beta H}\right\}}. \tag{3.7}$$

The ensemble average of n_κ is represented by the *Fermi distribution function* f_F:

$$\boxed{\langle n_\kappa\rangle \equiv \frac{\mathrm{TR}\{n_\kappa e^{\alpha N - \beta H}\}}{\mathrm{TR}\{e^{\alpha N - \beta H}\}} = [\exp(\beta\epsilon_\kappa - \alpha) + 1]^{-1} \equiv f_F(\epsilon_\kappa)} \tag{3.8}$$

(Problem 3.1.1). The parameter α in this expression is determined from

$$n = \frac{\langle N\rangle}{V} = \frac{1}{V}\sum_\kappa \frac{1}{\exp(\beta\epsilon_\kappa - \alpha) + 1} \equiv \frac{1}{V}\sum_\kappa f_F(\epsilon_\kappa). \tag{3.9}$$

Hereafter we drop the subscript F on f_F.

We now investigate the behavior of the Fermi distribution function $f(\epsilon)$ at very low temperatures. Let us set

$$\alpha \equiv \beta\mu \equiv \frac{\mu}{k_B T}. \tag{3.10}$$

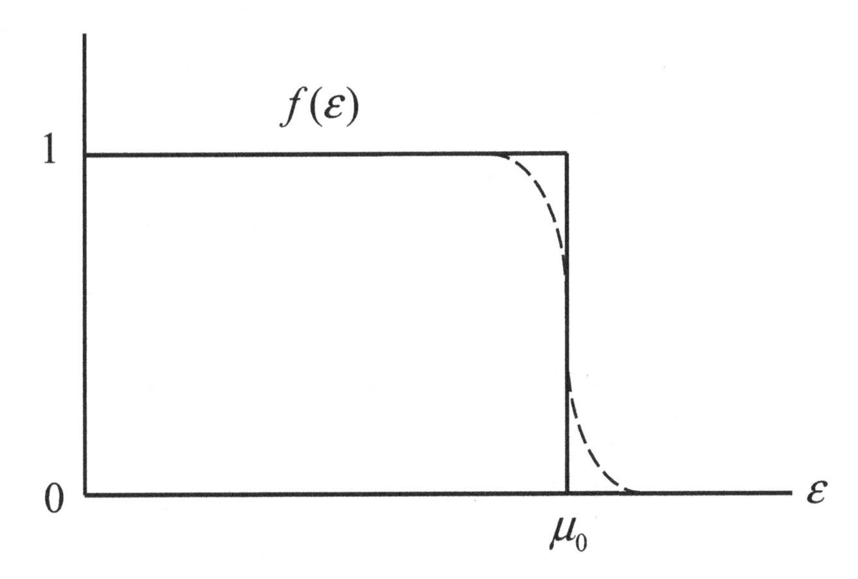

Figure 3.1: The Fermi distribution function against energy ϵ. The solid line is for $T = 0$, and the broken line is for a small T.

Here the quantity μ represents the *chemical potential*. In the low-temperature limit, the chemical potential μ approaches a positive constant $\mu_0 : \mu \to \mu_0 > 0$. We plot the Fermi distribution function $f(\epsilon)$ at $T = 0$ against the energy ϵ with a solid curve in Figure 3.1. It is a step function with the step at $\epsilon = \mu_0$. This means that every momentum state $\mathbf{p}_\kappa$ for which $\epsilon_\kappa = p_\kappa^2/2m < \mu_0$ is occupied with probability 1, and all other states are unoccupied. This special energy μ_0 is called the *Fermi energy*; it is often denoted by ϵ_F, which can be calculated as follows:

From Equation (3.9) we have

$$n = V^{-1} \sum_\kappa [f(\epsilon_\kappa)]_{T=0} = V^{-1} \times (\text{number of states } \kappa \text{ for which } \epsilon_\kappa \leq \mu_0). \tag{3.11}$$

The momentum eigenstates in Equation (3.2) can be represented by points in the three-dimensional momentum space, as shown in Figure 3.2. These points form a SC lattice with the lattice constant $2\pi\hbar/L$. Let us define the

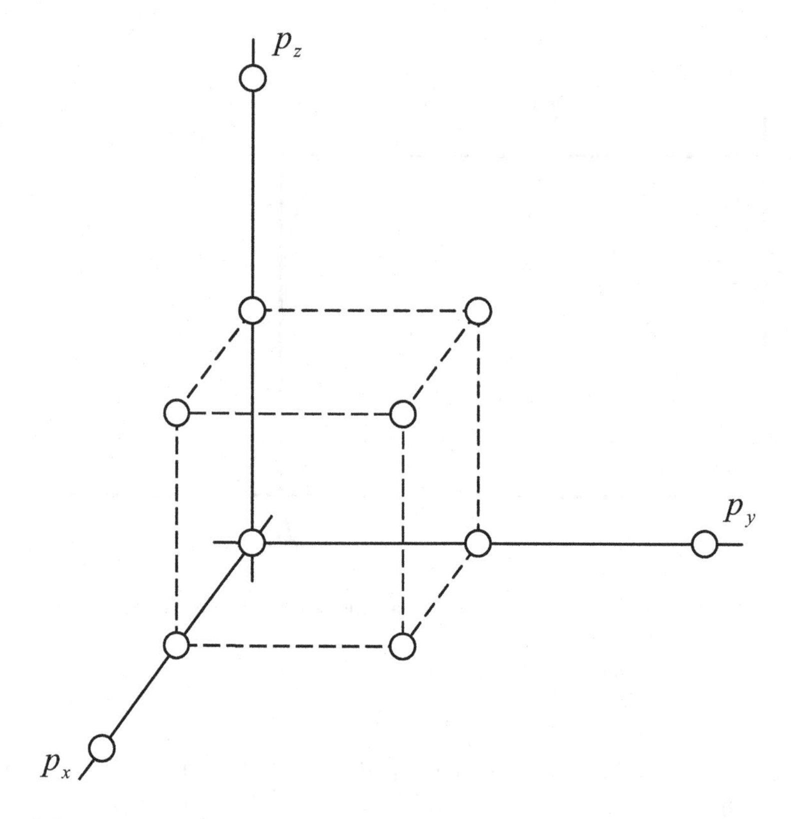

Figure 3.2: The momentum states for a periodic cubic boundary condition form a simple cubic lattice with the lattice constant $2\pi\hbar/L$.

Fermi momentum p_F by

$$\mu_0 = \epsilon_F \equiv \frac{p_F^2}{2m}. \tag{3.12}$$

The number of occupied states will be equal to the number of lattice points within the sphere of radius p_F. One lattice point corresponds to one unit cell for the SC lattice. Therefore, this number is equal to the volume of the sphere, $(4\pi/3)p_F^3$, divided by the volume of the unit cell, $(2\pi\hbar/L)^3$:

$$\text{Number of occupied states} = \left(\frac{4\pi}{3}\right)\frac{p_F^3}{(2\pi\hbar/L)^3}. \tag{3.13}$$

Introducing this into Equation (3.11), we obtain

$$n = \left(\frac{4\pi}{3}\right) \frac{p_F^3}{(2\pi\hbar/L)^3/L^3} = \left(\frac{4\pi}{3}\right) \frac{p_F^3}{(2\pi\hbar)^3}. \tag{3.14}$$

This result was obtained under the assumption of a periodic cube-box boundary condition. The result obtained in the *bulk limit*, where

$$L^3 = V \to \infty, \quad N \to \infty \quad \text{such that} \quad n \equiv \frac{N}{V} = \text{constant}, \tag{3.15}$$

is valid independent of the type of the boundary (see Problem 3.1.2).

In our discussion so far, we have neglected the fact that an electron has a *spin angular momentum* (or simply *spin*) as an additional degree of freedom. It is known that any quantum state for an electron must be characterized not only by the quantum numbers $(p_{x,j},\ p_{y,k},\ p_{z,l})$ describing its motion in the ordinary space but also by the quantum numbers describing its spin. It is further known that the electron has a permanent magnetic moment associated with its spin and that the eigenvalues s_z of the z-component of the electronic spin are discrete and restricted to the two values $\pm\hbar/2$. In the absence of a magnetic field, the magnetic potential energy is the same for both spin states. In the grand canonical ensemble, the states with the same energy are distributed with the same probability. In taking account of the spins, we must then multiply the right-hand side (RHS) of Equation (3.14) by the factor 2, called the *spin degeneracy factor*. We thus obtain

$$n = \frac{8\pi}{3} p_F^3 \frac{1}{(2\pi\hbar)^3} \tag{3.16}$$

(including the spin degeneracy). After solving this equation for p_F, we obtain the Fermi energy as follows:

$$\boxed{\epsilon_F = \frac{\hbar^2 (3\pi^2 n)^{2/3}}{2m}.} \tag{3.17}$$

Let us estimate the order of magnitude for ϵ_F by taking a typical metal Cu. This metal has a specific weight of 9 g cm^{-3} and a molecular weight of 63.5, yielding the number density $n = 8.4 \times 10^{22}$ electrons/cm^3. Using this value for n, we find that

$$\epsilon_F \equiv k_B T_F, \qquad T_F \cong 80,000 \text{ K}. \tag{3.18}$$

This T_F is called the *Fermi temperature*. The value found for the Fermi energy $\epsilon_F \equiv k_B T_F$ is very high compared to the thermal excitation energy of the order $k_B T$, which we shall see later in Section 3.3. This makes the thermodynamic behavior of the conduction electrons at room temperature drastically different from that of a classical gas.

The Fermi energy by definition is the chemical potential at 0 K. We may look at this relation in the following manner. For a box of a finite volume V, the momentum states form a SC lattice as shown in Figure 3.1. As the volume V is made greater, the unit-cell volume in the momentum space, $(2\pi\hbar/L)^3$, decreases like V^{-1}. However, we must increase the number of electrons N in proportion to V in the process of the bulk limit. Therefore, the radius of the *Fermi sphere* within which all momentum states are filled with electrons neither grows nor shrinks. Obviously this configuration corresponds to the lowest energy state for the system. The Fermi energy $\epsilon_F \equiv p_F^2/2m$ represents the electron energy at the surface of the Fermi sphere. If we attempt to add an extra electron to the Fermi sphere, we must bring in an electron with an energy equal to ϵ_F, meaning that $\epsilon_F = \mu_0$.

Problem 3.1.1. Verify Equation (3.8).

Problem 3.1.2. The momentum eigenvalues for a particle in a periodic rectangular box with sides of unequal lengths (L_1, L_2, L_3) are given by $p_{x,j} = 2\pi\hbar j/L_1$, $p_{y,k} = 2\pi\hbar k/L_2$, $p_{z,l} = 2\pi\hbar l/L_3$. Show that the Fermi energy ϵ_F for free electrons is still given as Equation (3.17) in the bulk limit.

3.2 Density of States

We must convert the *sum over quantum states* into an integral in many quantum statistical calculations. This conversion becomes necessary when we first find discrete quantum states for a periodic box, then seek the sum over states in the bulk limit. This conversion is a welcome procedure because the resulting integral is easier to handle than the sum. The conversion is purely mathematical in nature, but it is an important step in statistical mechanical computation.

Let us first examine a sum over momentum states corresponding to a one-dimensional motion. We take

$$\sum_k A(p_k), \tag{3.19}$$

where $A(p)$ is an arbitrary function of p. The discrete momentum states are equally spaced, as shown by the short bars in Figure 3.3. As the normalization length L increases, the spacing (distance) between two successive states $2\pi\hbar/L$ becomes smaller. This means that the number of states per unit momentum interval increases with L. We denote the number of states within a small momentum interval Δp by Δn. Consider the ratio $\Delta n/\Delta p$. Dividing both the numerator and denominator by Δp, we obtain

$$\frac{\Delta n}{\Delta p} = \frac{1}{\text{momentum spacing per state}} = \frac{L}{2\pi\hbar}. \tag{3.20}$$

This ratio $\Delta n/\Delta p$ increases linearly with the normalization length L.

Let us now consider a sum:

$$\sum_l A(p_l)\frac{\Delta n}{\Delta_l p}\Delta_l p, \tag{3.21}$$

where $\Delta_l p$ is the lth interval and p_l represents a typical value of p within the interval $\Delta_l p$, say the p-value at the midpoint of $\Delta_l p$. The two sums Equations (3.19) and (3.21) have the same dimension, and they are close to each other if (1) the function $A(p)$ is a smooth function of p and (2) there exist many states in $\Delta_l p$ so that $\Delta n/\Delta_l p$ can be regarded as the *density of states*. The second condition is satisfied for the momentum states $\{p_k\}$ when the length L is made sufficiently large. In the bulk limit, Equations (3.19) and (3.21) will be equal:

$$\lim_{L\to\infty} \sum_{k(\text{states})} A(p_k) = \sum_{\Delta_l p} A(p)\frac{\Delta n}{\Delta_l p}\Delta_l p. \tag{3.22}$$

In the small interval limit the sum on the RHS becomes the integral $\int dp\, A(p)\, dn/dp$, where [using Equation (3.20)]

$$\frac{dn}{dp} \equiv \lim_{\Delta p\to 0}\frac{\Delta n}{\Delta p} = \frac{L}{2\pi\hbar} \tag{3.23}$$

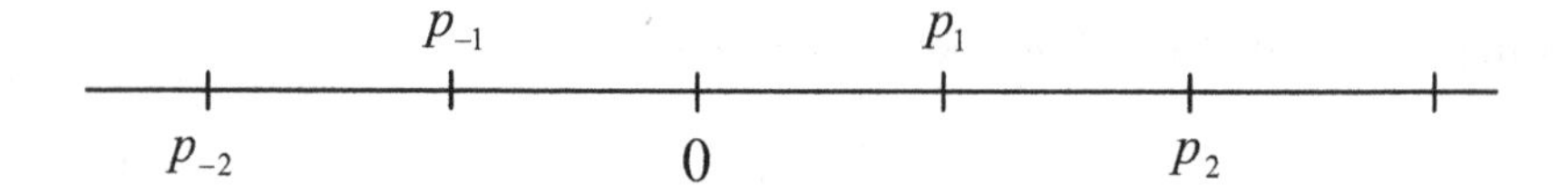

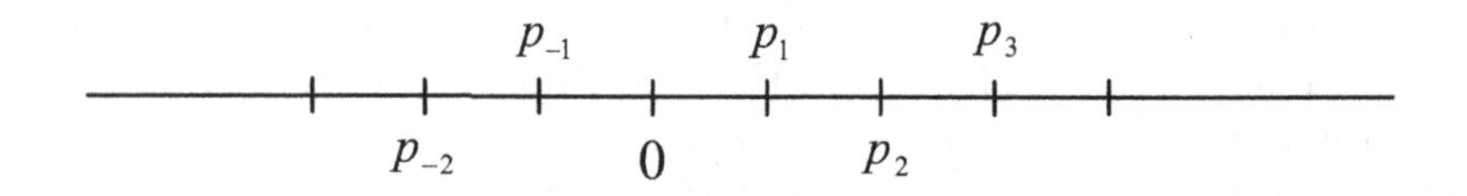

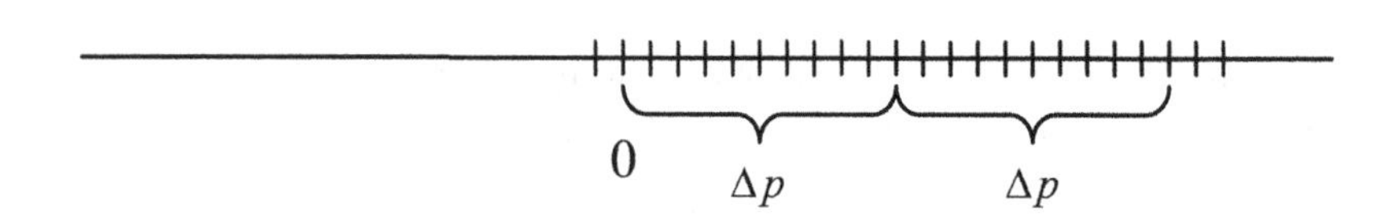

Figure 3.3: The linear momentum states are represented by short bars forming a linear lattice with unit spacing equal to $2\pi\hbar/L$.

is the *density of states* in the momentum space (line). In summary we obtain

$$\boxed{\sum_k A(p_k) \to \int_{-\infty}^{\infty} dp\, A(p)\frac{dn}{dp}.} \tag{3.24}$$

We stress that condition (1) depends on the nature of the function A. If $A(p)$ is singular at some point, then this condition is not satisfied, which may invalidate the limit in Equation (3.24). Such exceptional cases do occur in the Bose–Einstein condensation. We further note that the density of states in momentum $dn/dp = L(2\pi\hbar)^{-1}$ does not depend on the momentum.

The sum-to-integral conversion, which we have discussed, can easily be generalized for a multidimensional case. For example, we have

$$\sum_{\mathbf{p}_k} A(\mathbf{p}_k) \to \int d^3p\, A(\mathbf{p})\mathcal{D}(\mathbf{p}) \qquad \text{as } V \equiv L^3 \to \infty. \tag{3.25}$$

The density of states $\mathcal{D}(\mathbf{p}) \equiv dn/d^3p$ can be calculated by extending the arguments leading to Equation (3.20). We choose the periodic cubic box of side length L for the normalization, take the spin degeneracy into account, and obtain

$$\mathcal{D}(p) \equiv \frac{dn}{d^3p} = \frac{2L^3}{(2\pi\hbar)^3} \tag{3.26}$$

(with spin degeneracy).

Let us use this result and simplify the normalization condition in Equation (3.9). We obtain

$$\begin{aligned} n &= \mathrm{Lim}\frac{1}{V}\int d^3p\, f\left(\frac{p^2}{2m}\right)\mathcal{N}(\mathbf{p}) \\ &= 2(2\pi\hbar)^{-3}\int d^3p\, f\left(\frac{p^2}{2m}\right) \qquad (L^3 = V). \end{aligned} \tag{3.27}$$

Next consider the energy density of the system. Using Equations (3.5) and (3.8), we obtain

$$\begin{aligned} e \equiv \mathrm{Lim}\frac{\langle H\rangle}{V} &= \mathrm{Lim}\frac{1}{V}\sum_{\kappa} \epsilon_\kappa f(\epsilon_\kappa) \\ &= 2(2\pi\hbar)^{-3}\int d^3p \left(\frac{p^2}{2m}\right) f\left(\frac{p^2}{2m}\right). \end{aligned} \tag{3.28}$$

Equations (3.27) and (3.28) were obtained by starting with the momentum eigenvalues corresponding to the periodic cube-box boundary conditions. The results in the bulk limit, however, do not depend on the type of the boundary condition.

The concept of the density of states can also be applied to the energy domain. This is convenient when the sum over states has the form:

$$\sum_{\kappa} g(\epsilon_\kappa), \tag{3.29}$$

where $g(\epsilon_\kappa)$ is a function of the energy ϵ_κ associated with the state κ. The sums in Equations (3.5) and (3.9) are precisely in this form.

Let dn be the number of the states within the energy interval $d\epsilon$. In the bulk limit, this number dn will be proportional to the interval $d\epsilon$, so that

$$dn = \mathcal{N}(\epsilon)\, d\epsilon. \tag{3.30}$$

Here the proportionality factor

$$\mathcal{N}(\epsilon) \equiv \frac{dn}{d\epsilon} \tag{3.31}$$

is called the *density of states in the energy domain.* This quantity $\mathcal{N}(\epsilon)$ generally depends on the location of the interval $d\epsilon$. We may take the location to be the midpoint of the interval $d\epsilon$. If the set of the states $\{\kappa\}$ becomes densely populated in the bulk limit and the function g is smooth, then the sum may be converted into an integral of the form:

$$\boxed{\sum_{\kappa(\text{states})} g(\epsilon_\kappa) \to \int d\epsilon\, g(\epsilon)\mathcal{N}(\epsilon).} \tag{3.32}$$

Let us now calculate the density of states $\mathcal{N}(\epsilon)$ for the system of free electrons. The number of states dn in the spherical shell in momentum space is obtained by dividing the volume of the shell $4\pi p^2 dp$ by the unit cell volume $(2\pi\hbar/L)^3$ and multiplying the result by the spin degeneracy factor 2:

$$dn = \frac{2 \times 4\pi p^2\, dp}{(2\pi\hbar/L)^3} = V\frac{8\pi p^2}{(2\pi\hbar)^3}dp. \tag{3.33}$$

Since $p = (2m\epsilon)^{1/2}$, we obtain

$$dp = \frac{dp}{d\epsilon}d\epsilon = \left(\frac{m}{2\epsilon}\right)^{1/2} d\epsilon. \tag{3.34}$$

Using these equations, we obtain

$$dn = V\frac{8\pi(2m\epsilon)}{(2\pi\hbar)^3}\left(\frac{m}{2\epsilon}\right)^{1/2} d\epsilon = V\frac{8\pi 2^{1/2} m^{3/2}}{(2\pi\hbar)^3}\epsilon^{1/2} d\epsilon \tag{3.35}$$

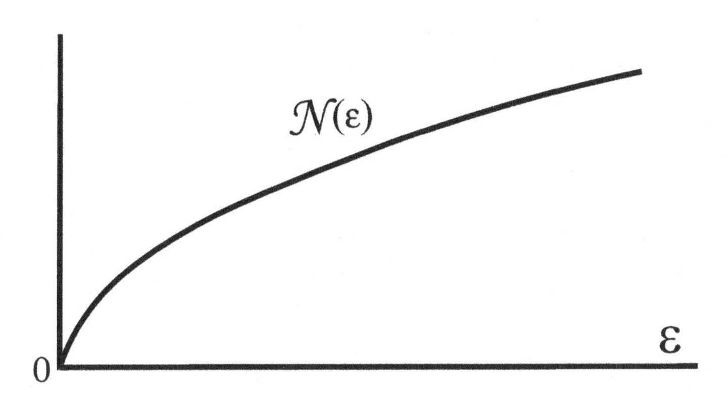

Figure 3.4: The density of states in energy, $\mathcal{N}(\epsilon)$, for free electrons in three dimensions grows like $\epsilon^{1/2}$.

or

$$\mathcal{N}(\epsilon) \equiv \frac{dn}{d\epsilon} = V\frac{2^{1/2}m^{3/2}}{\pi^2\hbar^3}\epsilon^{1/2}. \tag{3.36}$$

The density of states, $\mathcal{N}(\epsilon)$, grows like $\epsilon^{1/2}$ and is shown in Figure 3.4.

We may now restate the normalization condition (3.9) as follows:

$$n = \mathrm{Lim}\frac{1}{V}\sum_{\kappa} f(\epsilon_\kappa) = \mathrm{Lim}\frac{1}{V}\int_0^\infty d\epsilon\, f(\epsilon)\mathcal{N}(\epsilon) = \frac{2^{1/2}m^{3/2}}{\pi^2\hbar^3}\int_0^\infty d\epsilon\, \epsilon^{1/2} f(\epsilon). \tag{3.37}$$

The density of states in the energy domain, defined by Equation (3.31), is valid even when we have states other than the momentum states. We shall see such cases in later applications.

Problem 3.2.1. Obtain Equation (3.37) directly from Equation (3.27) by using the spherical polar coordinates $(p,\, \theta,\, \phi)$ in the momentum space, integrating over the angles $(\theta,\, \phi)$, and rewriting the p-integral in terms of the ϵ-integral.

3.3 Qualitative Discussions

At room temperature most metals have molar heat capacities of about $3R$ (where R is the gas constant) like nonmetallic solids. This experimental fact cannot be explained based on classical statistical mechanics. By applying the Fermi–Dirac statistics to conduction electrons, we can demonstrate the near absence of the electronic contribution to heat capacity. In this section, we shall show this in a qualitative manner.

Let us consider highly degenerate electrons with a high Fermi temperature T_F ($\approx 80,000$ K). At 0 K the Fermi distribution function:

$$f(\epsilon;\, \mu,\, T) \equiv \frac{1}{\exp\left[(\epsilon - \mu)/k_B T\right] + 1} \tag{3.38}$$

is a step function, as indicated by the dotted line in the lower diagram in Figure 3.5. At a finite temperature T, the abrupt drop at $\epsilon = \mu_0$ becomes a smooth drop, as indicated by a solid line in the same diagram. In fact, the change in the distribution function $f(\epsilon)$ is appreciable only in the neighborhood of $\epsilon = \mu_0$. The function $f(\epsilon;\, \mu,\, T)$ will drop from $1/2$ at $\epsilon = \mu$ to $1/101$ at $\epsilon - \mu = k_B T \ln(100)$ [which can be directly verified from Equation (3.38)]. This value $k_B T \ln(100) = 4.6\, k_B T$ is much less than the Fermi energy $\mu_0 = k_B T_F$ (80,000 K). This means that only those electrons with energies close to the Fermi energy μ_0 are excited by the rise in temperature. In other words, the electrons with energies ϵ far below μ_0 are not affected. There are many such electrons, and in fact this group of unaffected electrons forms the great majority.

The number N_X of electrons that are thermally excited can be found in the following manner. The density of states $\mathcal{N}(\epsilon)$ is shown in the upper diagram in Figure 3.5. Since $\mathcal{N}(\epsilon)\, d\epsilon$ represents by definition the number of electrons within $d\epsilon$, the integral of $\mathcal{N}(\epsilon)\, d\epsilon$ over the interval in which the electron population is affected gives an approximate number of excited electrons N_X. This integral can be represented by the shaded area in the upper diagram in Figure 3.5. Since we know from the earlier arguments that the affected range of the energy is of the order of $k_B T$ ($\ll \mu_0$), we can estimate N_X as

$$N_X = \text{shaded area in the upper diagram} \cong \mathcal{N}(\mu_0) k_B T \tag{3.39}$$

where $\mathcal{N}(\mu_0)$ is the density of states at $\epsilon = \mu_0$. From Equations (3.36) and

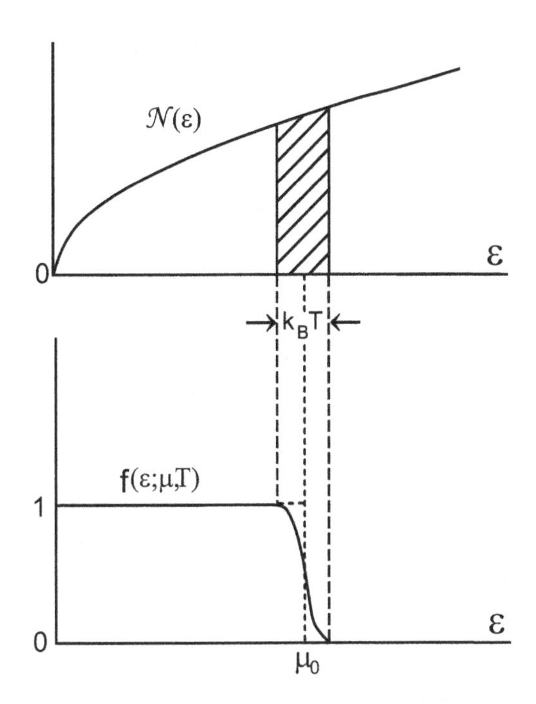

Figure 3.5: The density of states in energy, $\mathcal{N}(\epsilon)$, and the Fermi distribution function $f(\epsilon)$ are drawn as functions of the kinetic energy ϵ. The change in f is appreciable only near the Fermi energy μ_0 if $k_BT \ll \mu_0$. The shaded area represents the approximate number of thermally excited electrons.

(3.16), we obtain

$$\mathcal{N}(\mu_0) = V\frac{2^{1/2}m^{3/2}}{\pi^2\hbar^3}\mu_0^{1/2} = \frac{3N}{2\mu_0}. \tag{3.40}$$

Using this expression, we obtain from Equation (3.39)

$$N_X = \left(\frac{3}{2}\right)\frac{Nk_BT}{\mu_0}. \tag{3.41}$$

The electrons affected will move up with the extra energy of the order of k_BT per particle. Therefore, the change in the total energy, ΔE, will be approximately

$$\Delta E = N_X \times k_B T = \left(\frac{3}{2}\right) \frac{N(k_B T)^2}{\mu_0}. \tag{3.42}$$

Differentiating this equation with respect to T, we obtain

$$C_V = \frac{\partial}{\partial T} \Delta E = 3N_0 k_B^2 \frac{T}{\mu_0} = 3R \frac{T}{T_F} \quad (\mu_0 \equiv k_B T_F,\ R \equiv N_0 k_B) \tag{3.43}$$

for the molar heat capacity.

This expression indicates that the molar electronic heat capacity at room temperature ($T = 300$ K) is indeed small:

$$C_V = 3R \frac{300}{80,000} = 0.011R.$$

It is stressed that Equation (3.43) was obtained because the number of thermally excited electrons N_X is much less than the total number of electrons N [see Equation (3.41)]. We also note that the *electronic heat capacity is linear in the temperature.*

3.4 Quantitative Calculations

Historically Sommerfeld [1] first applied the Fermi–Dirac statistics to the conduction electrons and calculated the electronic heat capacity. His calculations resolved the heat capacity paradox (the absence of the electronic contribution). In this section we calculate the heat capacity quantitatively.

The heat capacity at constant volume, C_V, can be calculated by differentiating the internal energy E with respect to the temperature T:

$$C_V = \left.\frac{\partial E}{\partial T}\right|_V. \tag{3.44}$$

The internal energy density for free electrons given by Equation (3.28) can be expressed as

$$\frac{E(T,\,V)}{V} = \frac{2^{1/2} m^{3/2}}{\pi^2 \hbar^3} \int_0^\infty d\epsilon\, \epsilon^{3/2} f(\epsilon;\, \mu,\, T). \tag{3.45}$$

Here the chemical potential μ is related to the number density n by Equation (3.37):

$$n = \frac{2^{1/2} m^{3/2}}{\pi^2 \hbar^3} \int_0^\infty d\epsilon \, \epsilon^{1/2} f(\epsilon). \tag{3.46}$$

The integrals on the RHS of Equations (3.45) and (3.46) may be evaluated as follows. Let us assume

$$\alpha \equiv \beta\mu \gg 1. \tag{3.47}$$

Next we consider

$$F(x) \equiv \frac{1}{e^{x-\alpha}+1}, \qquad -\frac{dF}{dx} \equiv -F'(x) = \frac{e^{x-\alpha}}{(e^{x-\alpha}+1)^2}, \tag{3.48}$$

whose behaviors are shown in Figure 3.6. We note that $-dF/dx$ is a sharply peaked function near $x = \alpha$. Let us take the integral

$$I \equiv \int_0^\infty dx \, F(x) \frac{d\phi(x)}{dx}, \tag{3.49}$$

where $\phi(x)$ is a certain function of x. By integrating by parts, we obtain

$$\begin{aligned} I &= F(\infty)\phi(\infty) - F(0)\phi(0) - \int_0^\infty dx \, \phi(x) \frac{dF(x)}{dx} \\ &= -\phi(0) - \int_0^\infty dx \, \phi(x) \frac{dF(x)}{dx}, \end{aligned} \tag{3.50}$$

using Equations (3.47) and (3.48). We expand the function $\phi(x)$ into a Taylor series at $x = \alpha$,

$$\phi(x) = \phi(\alpha) + (x-\alpha)\phi'(\alpha) + \frac{1}{2}(x-\alpha)^2 \phi''(\alpha) + \cdots, \tag{3.51}$$

introduce it into Equation (3.50), and then integrate term by term. If $\alpha \gg 1$, then we obtain

$$\begin{aligned} -\int_0^\infty dx \, (x-\alpha)^n \frac{dF(x)}{dx} &= \int_{-\alpha}^\infty dy \, y^n \frac{e^y}{(e^y+1)^2} \\ &\cong \int_{-\infty}^\infty dy \, y^n \frac{e^y}{(e^y+1)^2} \equiv J_n. \end{aligned} \tag{3.52}$$

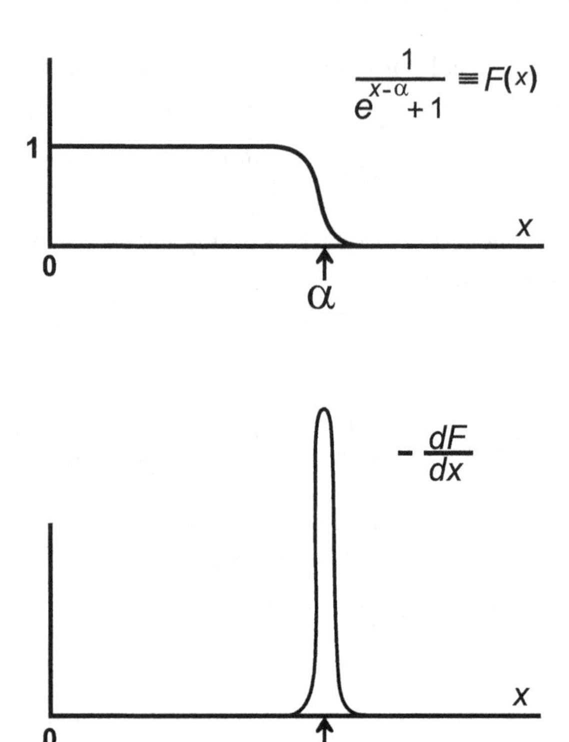

Figure 3.6: The functions $F(x)$ and $-dF/dx$ are shown for $x \geq 0$ and $\alpha \gg 1$.

The definite integrals J_n vanish for odd n since the integrands are odd functions of y. For small even n, we obtain

$$J_0 = 1,$$

$$J_2 = -2\int_{-\infty}^{\infty} dy\, y^2 \frac{d}{dy}\frac{1}{e^y+1} = -2\int_0^1 dz \frac{\ln z}{z+1} = \frac{\pi^2}{3}. \tag{3.53}$$

Using Equatios (3.49) through (3.53), we obtain

$$I = \int_0^{\infty} dx\, F(x)\frac{d\phi(x)}{dx} = \phi(\alpha) - \phi(0) + \frac{\pi^2}{6}\phi''(\alpha). \tag{3.54}$$

Equation (3.54) is useful if $\alpha \gg 1$ and $\phi(x)$ is a slowly varying function at $x = \alpha$.

We apply Equation (3.54) to evaluate the ϵ-integral in Equation (3.46)

$$\int_0^\infty d\epsilon\, \epsilon^{1/2} f(\epsilon) = \beta^{3/2} \int_0^\infty dx\, x^{3/2} F(x) \qquad (\beta\epsilon \equiv x).$$

We choose

$$\phi(x) = \frac{2}{3} x^{3/2}, \quad \frac{d\phi(x)}{dx} = x^{1/2}, \quad \frac{d^2\phi(x)}{dx^2} = \frac{1}{2} x^{-1/2},$$

and obtain

$$\begin{aligned} \int_0^\infty d\epsilon\, \epsilon^{3/2} f(\epsilon) &= \frac{2}{3\beta^{3/2}} (\beta\mu)^{3/2} + \frac{1}{\beta^{3/2}} \frac{\pi^2}{6} \frac{1}{2} (\beta\mu)^{-1/2} \\ &= \frac{2}{3} \mu^{3/2} \left(1 + \frac{\pi^2}{8} \beta^{-2} \mu^{-2}\right). \end{aligned} \qquad (3.55)$$

Using Equation (3.54), we obtain from Equation (3.46)

$$n = \frac{2^{3/2} m^{3/2}}{3\pi^2 \hbar^3} \mu^{3/2} \left(1 + \frac{\pi^2}{8} \beta^{-2} \mu^{-2}\right). \qquad (3.56)$$

Similary we can calculate the ϵ-integral in Equation (3.45) and obtain

$$e \equiv \frac{E(T, V)}{V} = \frac{2^{3/2} m^{3/2}}{5\pi^2 \hbar^3} \mu^{5/2} \left(1 + \frac{5\pi^2}{8} \beta^{-2} \mu^{-2}\right). \qquad (3.57)$$

The chemical potential μ, in general, depends on density n and temperature T: $\mu = \mu(n, T)$. This relation can be obtained from Equation (3.56). If we substitute $\mu(n, T)$ so obtained into Equation (3.57), we can regard the internal energy density as a function of n and T. By subsequently differentiating this function with respect to T at a fixed n, we can obtain the heat capacity at constant volume. We may also calculate the heat capacity by taking another route. The Fermi energy μ_0 is given by $\hbar^2 (3\pi^2 n)^{2/3}/2m$ [see Equation (3.16)], which depends only on n. Solving Equation (3.56) for μ and expressing the result in terms of μ_0 and T, we obtain (Problem 3.4.1)

$$\mu = \mu_0 \left[1 - \frac{\pi^2}{12} \left(\frac{k_B T}{\mu_0}\right)^2 + \cdots\right]. \qquad (3.58)$$

Introducing this expression into Equation (3.57), we obtain (Problem 3.4.2)

$$e = \frac{3}{5} n \mu_0 \left[1 + \frac{5\pi^2}{12} \left(\frac{k_B T}{\mu_0} \right)^2 \right]. \tag{3.59}$$

Differentiating this expression with respect to T at constant μ_0, we obtain

$$\boxed{C_V = \left(\frac{1}{2} \right) \pi^2 N k_B \frac{T}{T_F},} \tag{3.60}$$

where $T_F \equiv \epsilon_F / k_B$ is the *Fermi temperature*. Here we see that the heat capacity C_V for degenerate electrons is greatly reduced by the factor T/T_F ($\ll 1$) compared with the ideal-gas heat capacity $3Nk_B/2$. Also note that the heat capacity changes linearly with temperature T. These findings agree with the results of our previous qualitative calculations in Section 3.3.

At normal experimental temperatures the lattice contribution to the heat capacity, which was discussed in Chapter 2, is much greater than the electronic contribution. Therefore the experimental verification of the linear-T law must be done at very low temperatures, where the contribution of the lattice vibration becomes negligible. In this low-temperature region, the measured molar heat capacity should rise linearly with temperature T. By comparing the slope with

$$\frac{C_V}{T} = \frac{\pi^2}{2} \frac{R}{T_F} \tag{3.61}$$

[using Equation (3.60)], we can find the Fermi temperature T_F numerically. Since this temperature T_F is related to the effective mass m^* by

$$k_B T_F = \left(\frac{\hbar^2}{2m^*} \right) (3\pi^2 n)^{2/3} \tag{3.62}$$

[using Equations (3.59) and (3.60)], we can obtain the numerical value for the effective mass m^* of the conduction electron. Other ways of finding the m^*-value are through the transport and optical properties of conductors, which will be discussed later.

Problem 3.4.1. Use Equation (3.56) to verify Equation (3.58). Hint: Assume $\mu = \mu_0[1 + A(k_B T/\mu_0)^2]$ and find the constant A.

Problem 3.4.2. Verify Equation (3.59).

Chapter 4

Electric Conduction and the Hall Effect

Kinetic theory of the conduction electron without and with a constant magnetic field is developed in this chapter. The Landau Levels (LLs) for an electron subject to a constant magnetic field **B** is derived. The LL is highly degenerate, and its degeneracy is $eBA/(2\pi\hbar)$, where A is the sample area perpendicular to **B**. The Hall effect measurements give information about the charge sign and the density of the current carriers.

4.1 Ohm's Law and Matthiessen's Rule

Let us consider a system of free electrons moving in a potential field of impurities, which act as scatterers. The impurities by assumption are distributed uniformly.

Under an applied electric field **E** pointed along the positive x-axis, a classical electron will move according to Newton's equation of motion:

$$m\frac{dv_x}{dt} = -eE \tag{4.1}$$

(in the absence of the impurity potential). This gives rise to a uniform acceleration. The linear change in the velocity along the direction of field is expressed as

$$v_x = -\left(\frac{e}{m}\right) Et + v_x^0, \tag{4.2}$$

where v_x^0 is the x-component of the initial velocity. For a free electron the velocity v_x increases indefinitely and leads to infinite conductivity.

In the presence of the impurities, this uniform acceleration will be interrupted by scattering. When the electron hits a scatterer (impurity), the velocity will suffer an abrupt change in direction and grow again following Equation (4.2) until the electron hits another scatterer. Let us denote the *average time between successive scatterings* or *mean free time* by τ_f. The average velocity $\langle v_x \rangle$ is then given by

$$\langle v_x \rangle = -\frac{e}{m} E \tau_f, \tag{4.3}$$

where we assume that the electron loses the memory of its preceding motion every time it hits a scatterer, and the average velocity after collision is zero:

$$\left\langle v_x^0 \right\rangle = 0. \tag{4.4}$$

The charge current density (average current per unit volume) j is given by

$$j = (\text{charge}) \times (\text{carrier density}) \times (\text{velocity}) = en \langle v_x \rangle = e^2 n \tau_f \frac{E}{m}, \tag{4.5}$$

where n is the density of electrons. According to *Ohm's law*, the current density j is proportional to the applied field E when this field is small:

$$j = \sigma E. \tag{4.6}$$

The proportionality factor σ is called the *electrical conductivity*. It represents the facility with which the current flows in response to the electric field. Comparing the last two equations, we obtain

$$\boxed{\sigma = \frac{e^2 n \tau_f}{m}.} \tag{4.7}$$

This equation is very useful in the qualitative discussion of the electrical transport phenomenon. The inverse mass-dependence law means that the ion contribution to the electric transport in an ionized gas will be smaller by at least three orders of magnitude than the electron contribution. Also note that the conductivity is higher if the carrier density is greater and if the mean free time is greater.

The inverse of the mean free time τ_f,

$$\Gamma \equiv \frac{1}{\tau_f}, \tag{4.8}$$

is called the *rate of collision* or the *relaxation rate*. The relaxation rate Γ represents the mean frequency with which the electron is scattered by impurities. The collision rate Γ is given by

$$\Gamma = n_I v A, \tag{4.9}$$

where n_I is the density of scatterers, v is the electron speed, and A is the scattering cross section.

If there is more than one kind of scatterer, then the rate of collision may be computed by the addition law:

$$\Gamma = n_1 v A_1 + n_2 v A_2 + \cdots \equiv \Gamma_1 + \Gamma_2 + \cdots . \tag{4.10}$$

This is often called *Matthiessen's rule*. The total rate of collision is the sum of collision rates computed separately for each kind of scatterer.

Historically and also in practice, the analysis of resistance (R) data for a conductor proceeds as follows: If the electrons are scattered by impurities and again by phonons (quanta of lattice vibrations), the total resistance is the sum of the resistances due to each cause of scattering:

$$R_{\text{total}} = R_{\text{impurity}} + R_{\text{phonon}}. \tag{4.11}$$

This is the original statement of Matthiessen's rule. The electron-phonon scattering depends on temperature because of the changing phonon population, while the effect of the electron-impurity scattering is temperature independent. By separating the resistance in two parts, one temperature dependent and the other temperature independent, we may apply Matthiessen's rule. Since the resistance R is inversely proportional to the conductivity σ, Equations (4.7) and (4.10) together imply Equation (4.11).

Problem 4.1.1. Free electrons are confined within a long rectangular planar strip. Assume that each electron is *diffusely scattered* at the boundary so that it may move in all directions without preference after the scattering. Find the mean free path along the length of the strip. Calculate the conductivity.

Problem 4.1.2. Assume the same condition as in Problem 4.1.1 for the case in which electrons are confined within a long circular cylinder. Find the conductivity.

4.2 Motion of a Charged Particle in Electromagnetic Fields

Let us consider a particle of mass m and charge q moving in given electric and magnetic fields $(\mathbf{E}, \mathbf{B})$. In this section we shall study the motion of a charged particle, first classically and then quantum mechanically. We are mainly interested in situations where the electric field is very small in magnitude and the magnetic field may be arbitrarily large but constant in space and time.

Let us first consider the case in which $\mathbf{E} = 0$. Newton's equation of motion for a classical particle having a charge q in the presence of a magnetic field $\mathbf{B}$ is

$$m\frac{d\mathbf{v}}{dt} = q(\mathbf{v} \times \mathbf{B}). \tag{4.12}$$

We take the dot product of this equation with $\mathbf{v}$: $m\mathbf{v} \cdot d\mathbf{v}/dt = q\mathbf{v} \cdot (\mathbf{v} \times \mathbf{B})$. The RHS vanishes since $\mathbf{v} \cdot (\mathbf{v} \times \mathbf{B}) = (\mathbf{v} \times \mathbf{v}) \cdot \mathbf{B} = 0$. We obtain

$$m\mathbf{v} \cdot \frac{d\mathbf{v}}{dt} = \frac{d}{dt}\left(\frac{1}{2}mv^2\right) = 0, \tag{4.13}$$

which means that the *kinetic energy* $mv^2/2$ *is conserved.* This result is valid regardless of how the magnetic field $\mathbf{B}$ varies in space. If the magnetic field $\mathbf{B}$ varies in time, an electric field is necessarily induced, and the energy will not be conserved.

In the case of a constant magnetic field, we can rewrite Equation (4.12) as

$$\frac{d\mathbf{v}}{dt} = \mathbf{v} \times \boldsymbol{\omega}_c, \tag{4.14}$$

where $\boldsymbol{\omega}_c$ is the constant vector pointing along the direction of the magnetic field and having the magnitude:

$$\omega_c \equiv \frac{qB}{m}. \tag{4.15}$$

This quantity ω_c is called the *cyclotron frequency*. It is proportional to the magnetic field strength and inversely proportional to the mass of the particle. For example, for an electron in a field of 1000 gauss, we have $\omega_c \sim 10^{10}$ sec^{-1}.

From Equation (4.14) we deduce that the motion of the electron consists of the uniform motion along the magnetic field with velocity v_z plus a circular motion with constant speed $v_\perp$ about the magnetic field. The radius R of this circular orbit about the magnetic field, called the *cyclotron radius*, can be determined from centripetal force = magnetic force:

$$\frac{mv_\perp^2}{R} = ev_\perp B. \tag{4.16}$$

Solving this equation for R, we obtain

$$R = \frac{mv_\perp^2}{eB} = \frac{v_\perp}{\omega_c}. \tag{4.17}$$

If $B = 1000$ gauss and $v_\perp = 10^5$ cm/sec, the cyclotron radius R is on the order of 10^{-5} cm = 1000 Å. Note: The radius is inversely proportional to B. Thus as the magnetic field becomes greater, the electron spirals more rapidly in smaller circles.

As we saw earlier, the magnetic field $\mathbf{B}$ cannot change the kinetic energy of the electron. The cyclotron motion describes an *orbit of constant energy*. This feature is preserved in quantum mechanics. In fact it can be used to explore the Fermi (constant-energy) surface for the conduction electrons in a metal.

The motion of an electron in static, uniform electric and magnetic fields is similar to the case we have just discussed. First, if the electric field $\mathbf{E}$ is applied along the direction of $\mathbf{B}$, the motion perpendicular to $\mathbf{B}$ is not affected so that the electron spirals around the magnetic field lines. The motion along $\mathbf{B}$ is subjected to a uniform acceleration equal to $-e\mathbf{E}/m$.

Let us now turn to the second and more interesing situation in which electric and magnetic fields are perpendicular to each other. The drift velocity $\mathbf{v}_D$ defined by

$$\mathbf{v}_D \equiv \frac{\mathbf{E} \times \mathbf{B}}{B^2} \tag{4.18}$$

is constant in time and has the dimensions of velocity. Let us decompose the velocity $\mathbf{v}$ in two parts:

$$\mathbf{v} \equiv \mathbf{v}' + \mathbf{v}_D. \tag{4.19}$$

Substituting this into the equation of motion

$$m\frac{d\mathbf{v}}{dt} = \text{Lorentz force} = q(\mathbf{E} + \mathbf{v} \times \mathbf{B}), \tag{4.20}$$

we obtain

$$\text{Left-hand side (LHS)} = m\frac{d}{dt}(\mathbf{v}' + \mathbf{v}_D) = m\frac{d\mathbf{v}'}{dt}$$

and

$$\begin{aligned} \text{RHS} &= q(\mathbf{E} + \mathbf{v}_D \times \mathbf{B} + \mathbf{v}' \times \mathbf{B}) \\ &= q\left[\mathbf{E} + \frac{(\mathbf{E} \times \mathbf{B}) \times \mathbf{B}}{B^2}\right] + q(\mathbf{v}' \times \mathbf{B}) = q(\mathbf{v}' \times \mathbf{B}) \end{aligned}$$

or

$$m\frac{d\mathbf{v}'}{dt} = q(\mathbf{v}' \times \mathbf{B}), \tag{4.21}$$

which has same form as Equation (4.12). The motion can then be regarded as the superposition of the motion in a uniform magnetic field and a drift of the cyclotron orbit with the constant velocity v_D as given in Equation (4.18). This motion is shown in Figure 4.1.

The *drift velocity* $\mathbf{v}_D$ in Equation (4.18) is perpendicular to both $\mathbf{E}$ and $\mathbf{B}$. This implies that the weak electric field will induce a macroscopic current $\mathbf{j}$ in the direction perpendicular to both $\mathbf{E}$ and $\mathbf{B}$:

$$\mathbf{j} = qn\frac{(\mathbf{E} \times \mathbf{B})}{B^2}, \tag{4.22}$$

where n is the number density of the electrons. This current is called the *Hall current*. We note that drift velocity v_D is independent of charge and mass, which is significant. This turns out to be a practically important property. The measurement of the Hall effect gives information about the type of the charge carrier ("electron" or "hole") and the number density of carriers.

4.3 The Landau States and Levels

We have so far discussed the motion of an electron using classical mechanics. Most of the qualitative features also hold true in quantum mechanics. The

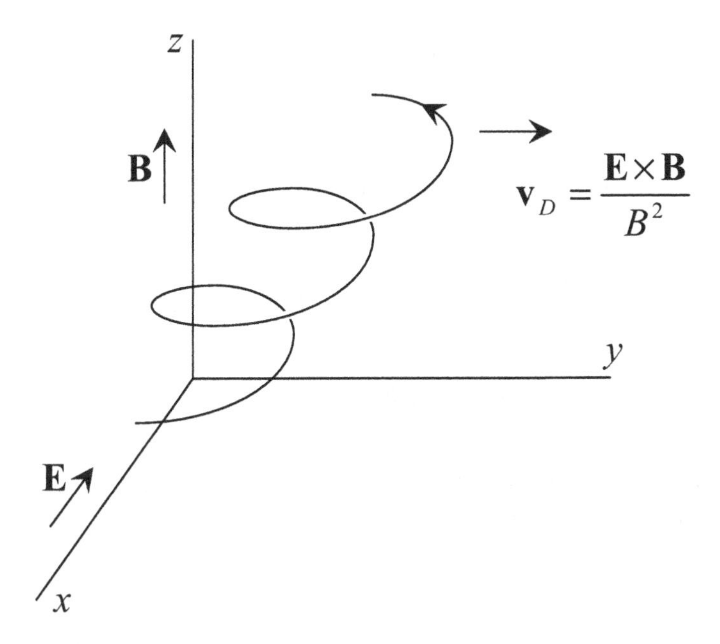

Figure 4.1: An "electron" in this figure spirals under the electric (**E**) and magnetic (**B**) fields, which are perpendicular to each other.

most important quantum effect is the *quantization of the cyclotron motion*. To see this, let us calculate the energy levels of an electron in a constant magnetic field **B**. We choose the *vector potential*

$$\mathbf{A} = (A_x,\, A_y,\, A_z) = (0,\, Bx,\, 0), \tag{4.23}$$

which yields a constant field **B** in the z direction (Problem 4.3.1). The Hamiltonian H then is given by

$$H = \frac{|\mathbf{p} + e\mathbf{A}|^2}{2m} = \frac{[p_x^2 + (p_y + eBx)^2 + p_z^2]}{2m}. \tag{4.24}$$

This Hamiltonian is derived in Appendix A. The Schrödinger equation can now be written down as

$$-\frac{\hbar^2}{2m}\left[\frac{\partial^2}{\partial x^2} + \left(\frac{\partial}{\partial y^2} + \frac{ieB}{\hbar}x\right)^2 + \frac{\partial^2}{\partial z^2}\right]\psi = E\psi. \tag{4.25}$$

Since the Hamiltonian H contains neither y nor z explicitly, we assume a wave function of the form

$$\psi(x, y, z) = e^{-i(k_y y + k_z z)} \phi(x). \tag{4.26}$$

Substituting this expression into Equation (4.25) yields the following equation for $\phi(x)$:

$$\left[-\frac{\hbar^2}{2m} \frac{d^2}{dx^2} + \frac{1}{2} m \omega_c^2 \left(x - \frac{\hbar k_y}{eB} \right)^2 \right] \phi(x) = E_1 \phi(x), \tag{4.27}$$

$$E_1 \equiv E - \frac{\hbar^2 k_z^2}{2m} \tag{4.28}$$

(Problem 4.3.3). Equation (4.27) is the energy eigenvalue equation for a harmonic oscillator with the angular frequency $\omega_c \equiv eB/m$ and the center of oscillation displaced from the origin by

$$X = \frac{\hbar k_y}{eB}. \tag{4.29}$$

The energy eigenvalues are given by

$$E_1 = \left(n + \frac{1}{2} \right) \hbar \omega_c, \qquad n = 0, 1, 2, \ldots. \tag{4.30}$$

Combining this with Equation (4.28), we obtain

$$\boxed{E = \left(n + \frac{1}{2} \right) \hbar \omega_c + \frac{\hbar^2 k_z^2}{2m}.} \tag{4.31}$$

These energy eigenvalues are called the *Landau Levels* (LLs) [1]. The corresponding quantum states, called the *Landau states*, are characterized by the quantum number (n, k_y, k_z). We note that the energies do not depend on k_y, and they are therefore highly degenerate. The Landau states are quite different from the momentum eigenstates. This has significant consequences on magnetization and galvanomagnetic phenomena. The electron in a Landau state may be pictured as in a circulation with the angular frequency ω_c around the magnetic field. If a radiation having a frequency equal to ω_c is applied, the electron may jump up from one Landau state to another by absorption of a photon energy equal to $\hbar\omega_c$. This generates a phenomenon of *cyclotron resonance*.

Problem 4.3.1. Show that the vector potential given by Equation (4.23) generates a magnetic field pointing in the positive z-direction.

Problem 4.3.2. For a constant magnetic field $\mathbf{B}$ we can choose the vector potential $\mathbf{A} = (1/2)\mathbf{B} \times \mathbf{r}$. Show this by explicitly calculating $\nabla \times \mathbf{A}$.

Problem 4.3.3. Derive Equation (4.27) from Equations (4.25) and (4.26).

4.4 The Degeneracy of the Landau Levels

The Landau levels are highly degenerate. The *degeneracy*, that is, the number of the electrons that can occupy each Landau level, is

$$\frac{eBA}{2\pi\hbar}, \tag{4.32}$$

where A is the sample area perpendicular to the magnetic field $\mathbf{B}$. This is shown in this section.

Let us consider a particle moving along a straight line of length L. In classical mechanics, a dynamical state of the system, (x, p), can be represented by a point in the phase space. In quantum mechanics, the dynamical state cannot be represented by a point because of the Heisenberg uncertainty principle. The set of momentum eigenstates,

$$p_k \equiv 2\pi\hbar k/L\,, \quad k = \ldots, -1, 0, 1, 2, \ldots, \tag{4.33}$$

however, may be represented by the set of quantum cells in phase space as shown in Figure 4.2. We note that

1. Each cell extends from 0 to L in position, implying that a particle can be found with an equal probability everywhere along the line from 0 to L.

2. The area of each cell equals $2\pi\hbar$.

Such representation of the quantum states in phase space is not restricted to this special case. The quantum states for a simple harmonic oscillator can be represented by the elliptical shells as shown in Figure 4.3. Each cell has an area equal to $2\pi\hbar$.

Let us now consider the case with a constant magnetic field $\mathbf{B}$. The Hamiltonian H is given by Equation (4.24). We can check (Problem 4.4.1)

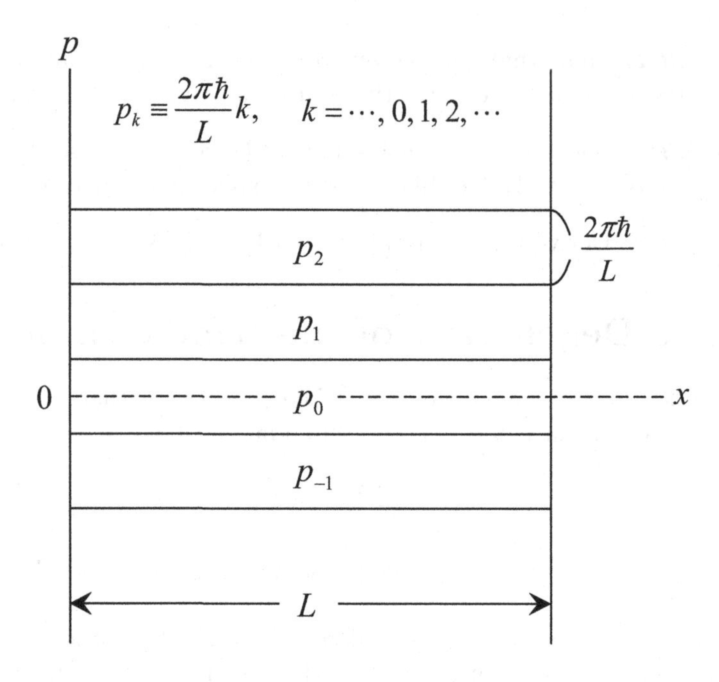

Figure 4.2: The momentum states $\{p_k\}$ for linear motion are represented by rectangular cells of equal area $(2\pi\hbar)$ in phase space.

that one set of the Hamilton's equations gives the correct Newton's equation of motion and the second set yields

$$m\dot{\mathbf{r}} \equiv m\mathbf{v} = \mathbf{p} + e\mathbf{A} \equiv \mathbf{\Pi}. \tag{4.34}$$

Note that this *kinetic momentum* $\mathbf{\Pi}$ is not equal to the canonical momentum $\mathbf{p}$. The component Π_z is equal to p_z but $(\Pi_x, \Pi_y) \neq (p_x, p_y)$. We now go over to quantum mechanics. Using the *quantum condition*

$$\begin{aligned} [x,\, p_x] &= [y,\, p_y] = i\hbar, \\ [x,\, y] &= [x,\, p_y] = [y,\, p_y] = \cdots = 0, \end{aligned} \tag{4.35}$$

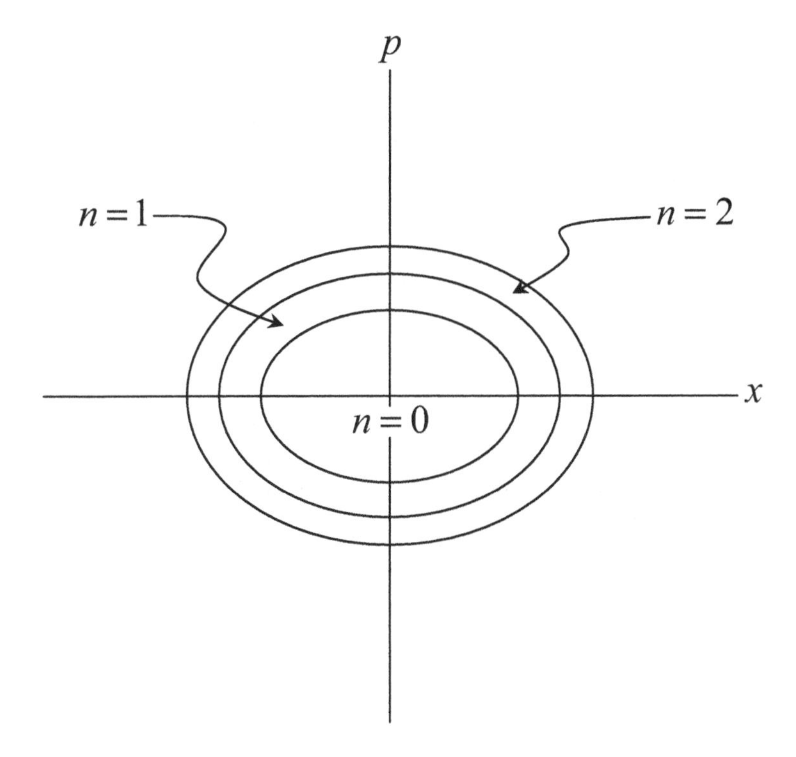

Figure 4.3: The quantum mechanical eigenstates for a simple harmonic oscillator are represented by the quantum cells of phase-space area $2\pi\hbar$.

we obtain (Problem 4.4.2)

$$
\begin{aligned}
[\Pi_x, \Pi_y] &= -ie\hbar B \, , \\
[\Pi_z, \Pi_x] &= [\Pi_y, \Pi_z] = 0.
\end{aligned}
\tag{4.36}
$$

The noncommutativity of Π_x and Π_y means that the x- and y-motion are correlated, making the motion one-dimensional (1D) harmonic oscillatorlike instead of 2D oscillatorlike.

In fact, we can illustrate this behavior in more detail as follows: We introduce new canonical variables:

$$\frac{1}{\sqrt{m}}\,\Pi_x = P\,, \quad \frac{1}{eB}\sqrt{m}\,\Pi_y = Q. \tag{4.37}$$

We can then rewrite part of the Hamiltonian as

$$H_\perp \equiv \frac{1}{2m}(\Pi_x^2 + \Pi_y^2) \equiv \frac{1}{2m}\Pi_\perp^2 = \frac{1}{2}P^2 + \frac{1}{2}\omega_c^2 Q^2. \tag{4.38}$$

Using Equations (4.36) and (4.37), we obtain

$$[Q, P] = i\hbar. \tag{4.39}$$

The last two equations mean that the energy eigenvalues $E_\perp$ are the same as those of the harmonic oscillator with the cyclotron frequency ω_c. Hence, we obtain

$$E_\perp = \left(N_L + \frac{1}{2}\right)\hbar\omega_c\,, \quad N_L = 0, 1, 2, \ldots, \tag{4.40}$$

in agreement with Equation (4.31).

After simple calculations, we can show (Problem 4.4.3) that

$$dx\, d\Pi_x\, dy\, d\Pi_y = dx\, dp_x\, dy\, dp_y. \tag{4.41}$$

We can now represent the circulational part of the quantum states by a small quasi-phase space cell of the volume $(2\pi\hbar)^{-2}dx\, d\Pi_x\, dy\, d\Pi_y$. The Hamiltonian $H_\perp$ in Equation (4.38) does not depend on the position $(x,\ y)$. Assuming large normalization lengths $(L_1,\ L_2)$, $A = L_1L_2$, we can then represent the partial Landau states by the concentric shells of the phase space volume

$$(2\pi)\,\Pi_\perp\,\Delta\Pi_\perp \cdot L_1L_2(2\pi\hbar)^{-2} = A(2\pi\hbar)^{-1}\omega_c m^* \tag{4.42}$$

with the energy separation

$$\hbar\omega_c = \Delta\left(\frac{\Pi_\perp^2}{2m^*}\right) = \frac{\Pi_\perp\,\Delta\Pi_\perp}{m^*}, \tag{4.43}$$

as shown in Figure 4.4. The number of quantum states, dN, between the neighboring orbits is given by the phase space volume over $(2\pi\hbar)^2$:

$$dN = \frac{L_1L_2}{(2\pi\hbar)^2}(2\pi)\Pi_\perp\Delta\Pi_\perp. \tag{4.44}$$

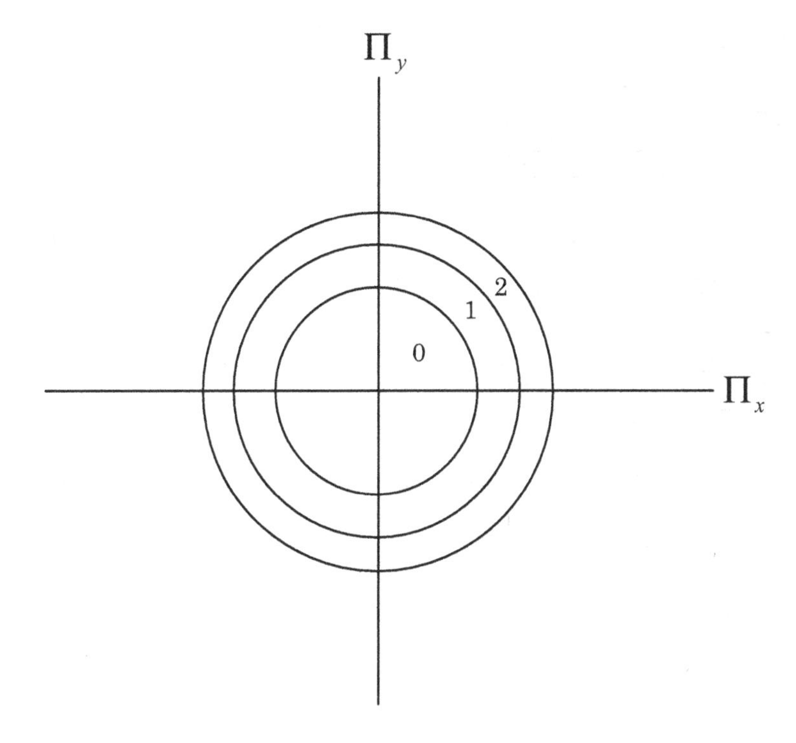

Figure 4.4: The circulational part of the Landau states is represented by the circular shells in the (Π_x, Π_y) space.

Using Equation (4.43), we obtain

$$dN = \frac{eBA}{2\pi\hbar}, \quad A = L_1 L_2, \tag{4.45}$$

establishing Equation (4.32).

In summary, as the field strength is raised, the LL separation $\hbar\omega_c$ increases. Hence, the area between the neighboring orbits $2\pi\Pi_\perp\Delta\Pi_\perp$ increases in proportion to $\hbar\omega_c$, and the degeneracy dN becomes greater. This behavior is shown in Figure 4.5.

Problem 4.4.1. Obtain Hamilton's equations of motion using the Hamiltonian given in Equation (4.24). Show that these equations are equivalent to Newton's equation of motion.

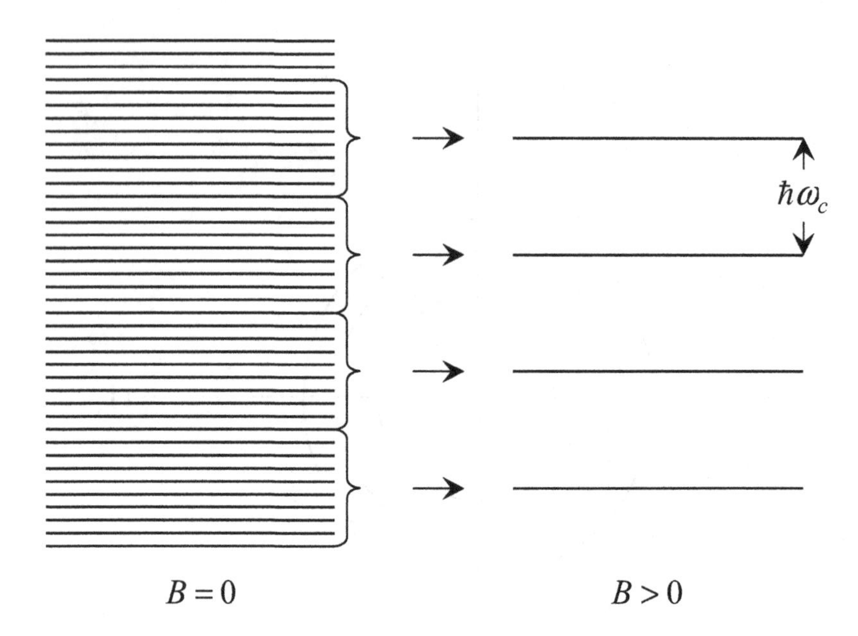

Figure 4.5: The equidistant energy levels for 2D free electrons are bundled into the Landau levels.

Problem 4.4.2. Verify Equation (4.36).

Problem 4.4.3. Verify Equation (4.44).

4.5 The Hall Effect: "Electrons" and "Holes"

In this section we will discuss the Hall effect. As we see later, "electrons" and "holes" play very important roles in the theory of conducting matter.

Let us consider a conducting wire connected to a battery. If a magnetic field $\mathbf{B}$ is applied, the field penetrates the wire. The *Lorentz force*

$$\mathbf{F} = q\mathbf{v} \times \mathbf{B} \tag{4.46}$$

acts on the classical electron. Here the picture of the straight-line motion of a free electron in kinetic theory fails to apply. When the field (magnitude) B is not too high and the stationary state is considered, the actual physical situation turns out to be much simpler.

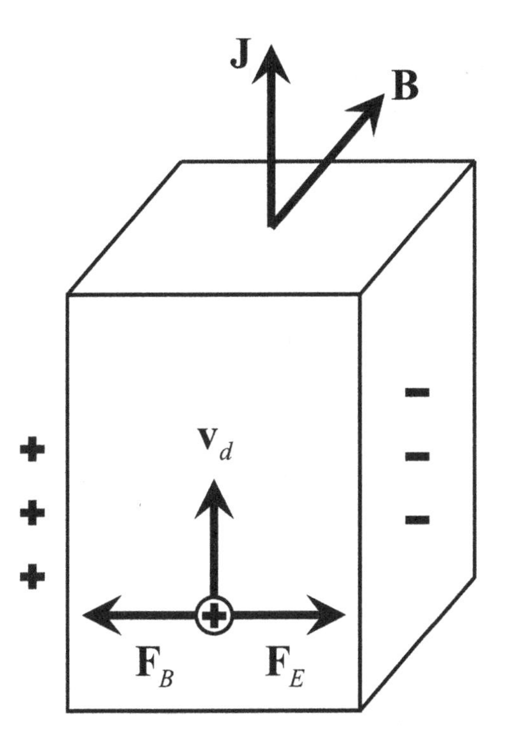

Figure 4.6: The magnetic and electric forces ($\mathbf{F}_B$, $\mathbf{F}_E$) are balanced to zero in the Hall effect measurement.

Take the case in which the field $\mathbf{B}$ is applied perpendicular to the wire of a rectangular cross section as shown in Figure 4.6. Experiments show that voltage V_H is generated perpendicular to the field $\mathbf{B}$ and the electric current $\mathbf{J}$ such that a steady current flows in the wire unhindered. We may interpret this condition as follows: Let us write the current density $\mathbf{j}$ as

$$\mathbf{j} = qn\mathbf{v}_d, \tag{4.47}$$

where n is the density of conduction electrons and $\mathbf{v}_d$ the *drift velocity*. A charge carrier having a velocity equal to the drift velocity $\mathbf{v}_d$ is subjected to the *Lorentz force*

$$\mathbf{F} = q(\mathbf{E}_H + \mathbf{v}_d \times \mathbf{B}), \tag{4.48}$$

Table 4.1: Hall Coefficients for Selected Metals

Metal	Valence	$-1/qnR_H$
Li	1	−0.8
Na	1	−1.2
K	1	−1.1
Cu	1	−1.5
Ag	1	−1.3
Au	1	−1.5
Be	2	0.2
Mg	2	0.4
In	3	0.3
Al	3	0.3

where $\mathbf{E}_H$ is the electric field due to the Hall voltage V_H. In the geometry shown in Figure 4.6 only the x-component of the force $\mathbf{F}$ is relevant. If the net force vanishes so that

$$E_H = v_d B \qquad \text{(magnitude)}, \tag{4.49}$$

the carrier can proceed along the z-direction unhindered.

Let us check our model calculation. We define the *Hall coefficient* R_H by

$$R_H \equiv \frac{E_H}{jB}, \tag{4.50}$$

where the three quantities (E_H, j, B) on the RHS can be measured. Using Equations (4.32) and (4.34), we obtain

$$\boxed{R_H = -\frac{1}{qn}.} \tag{4.51}$$

The experimental values of $-qnR_H$ for some metals are given in Table 4.1.

For alkali metals the agreement between theory and experiment is nearly perfect. This is quite remarkable. Equation (4.48) means a zero net balanace between the magnetic force $F_B = qv_dB$ and the electric force $F_E = qE_H$ only in the average sense through the drift velocity v_d. In real metals there must be a great number of electrons moving under unbalanced forces.

The measured Hall coefficients R_H for most metals are negative, meaning that the charge carrier are "electrons" having a negative charge $q = -e$. There are exceptions. As we see in Table 4.1, Al, Be, and others exhibit positive Hall coefficients. This means that there are charge carriers, called "holes," *having a positive charge* $q = +e$. This is the quantum many-body effect. As we shall see later, the existence of "electrons" and "holes" is closely connected with the curvature of the Fermi surface. Nonmagnetic metals that have "holes" tend to be superconductors.

Chapter 5

Magnetic Susceptibility

The electron has mass m, charge $-e$, and a half spin. Hence it has a spin magnetic moment. Pauli paramagnetism and Landau diamagnetism of the conduction electrons are discussed in this chapter.

5.1 The Magnetogyric Ratio

Let us consider a classical electron's motion in a circle in the xy-plane as shown in Figure 5.1. The angular momentum $\mathbf{j} \equiv \mathbf{r} \times \mathbf{p}$ points in the positive z-axis and its magnitude is given by

$$mrv. \tag{5.1}$$

According to the electromagnetic theory, a current loop generates a magnetic moment $\boldsymbol{\mu}$ (vector) whose magnitude equals the current times the area of the loop and whose direction is specified by the right-hand screw rule. The magnitude of the moment generated by the electron motion, therefore, is given by

$$\text{Current} \times \text{area} = \left(\frac{ev}{2\pi r}\right)(\pi r^2) = \frac{1}{2}evr, \tag{5.2}$$

and the direction is along the negative z-axis. We observe here that *magnetic moment* $\boldsymbol{\mu}$ *is proportional to the angular momentum* $\mathbf{j}$. We may express this relation by

$$\boxed{\boldsymbol{\mu} = \alpha\, \mathbf{j}.} \tag{5.3}$$

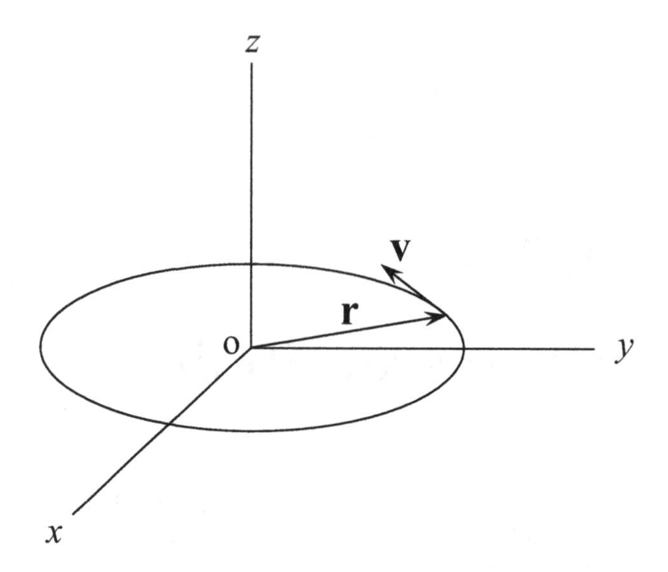

Figure 5.1: An electron in a circular motion generates a magnetic moment $\boldsymbol{\mu}$ proportional to its angular momentum $\mathbf{j}$.

This relation, in fact, holds not only for this circular motion but also in general. The proportionality factor

$$\alpha = \frac{-e}{2m} \tag{5.4}$$

is called the *magnetogyric* or *magnetomechanical ratio.* We note that the ratio is proportional to the charge $-e$ and inversely proportional to the mass m.

Let us assume that a magnetic field $\mathbf{B}$ is applied along the positive z-axis. The potential energy V of a magnetic dipole with moment $\boldsymbol{\mu}$ is given by

$$\boxed{V = -\mu B \cos\theta = -\mu_z B,} \tag{5.5}$$

where θ is the angle between the vectors $\boldsymbol{\mu}$ and $\mathbf{B}$.

We may expect that a general relation such as Equation (5.3) holds also in quantum theory. The angular momentum (eigenvalues) is quantized in the units of $\hbar$. The electron has a spin angular momentum $\mathbf{s}$ whose z-component can assume either $(1/2)\hbar$ or $-(1/2)\hbar$. Let us write

$$s'_z = \frac{1}{2}\hbar\sigma'_z \equiv \frac{1}{2}\hbar\sigma, \qquad \sigma \equiv \sigma'_z = \pm 1. \tag{5.6}$$

Analogous to Equation (5.3) we assume that

$$\mu_z \propto s'_z \propto \sigma. \tag{5.7}$$

We shall write this quantum relation in the form

$$\boxed{\mu_z = \frac{1}{2} g \mu_B \sigma,} \tag{5.8}$$

where

$$\mu_B \equiv \frac{e\hbar}{2mc} = 0.927 \times 10^{-20} \quad \text{erg gauss}^{-1}, \tag{5.9}$$

called the *Bohr magneton*, has the dimensions of magnetic moment. The constant g in Equation (5.8) is a numerical factor of order 1, and is called a *g-factor*. If the magnetic moment of the electron is accounted for by the "spinning" of the charge around a certain axis, then the g-factor should be exactly one. The experiments, however, show that this factor is 2. This phenomenon is known as the *spin anomaly*, which is an important indication of the quantum nature of the spin.

In the presence of a magnetic field **B** the electron whose spin is directed along **B**, defined as the electron with the *up spin*, will have a lower energy than the *down spin* electron whose spin is directed against **B**. The difference is, according to Equations (5.5) and (5.8),

$$\Delta\epsilon = \frac{1}{2} g\mu_B(+1) - \frac{1}{2} g\mu_B(-1) = g\mu_B B. \tag{5.10}$$

For $B = 7000$ gauss and $g = 2$, we obtain the numerical estimate $\Delta\epsilon/k_B \simeq$ 1 K.

If an electromagnetic wave with the frequency ν satisfying $h\nu = \Delta\epsilon$ is applied, then the electron may absorb a photon of the energy $h\nu$ and jump to the upper energy level. Figure 5.2 illustrates this phenomenon, which is known as the *electron spin resonance*. The frequency corresponding to $\Delta\epsilon/k_B = 1$ K is

$$\nu = 2.02 \times 10^{10} \text{ cycles sec}^{-1}. \tag{5.11}$$

This frequency falls in the microwave region of the electromagnetic radiation spectrum.

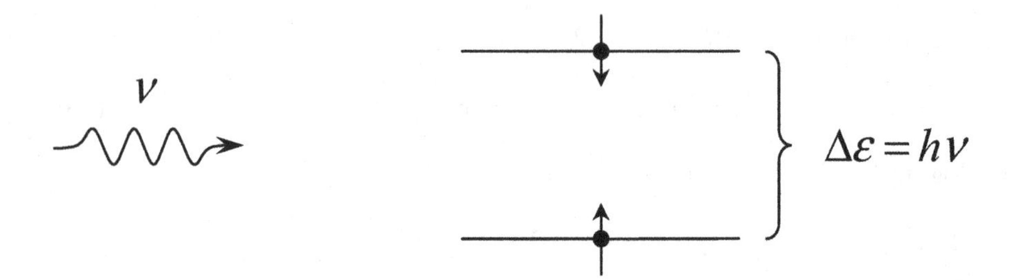

Figure 5.2: An electron with the up spin may absorb a photon of the energy $h\nu$ and jump to the upper energy level by flipping its spin if $\Delta\epsilon = h\nu$ (electron spin resonance).

5.2 Pauli Paramagnetism

Let us consider an electron moving in free space. The quantum states for the electron can be characterized by momentum $\mathbf{p}$ and spin σ $(= \pm 1)$. If a weak constant magnetic field $\mathbf{B}$ is applied along the positive z-axis, then the energy ϵ associated with the quantum state $(\mathbf{p}, \sigma)$ is given by

$$\epsilon = \frac{p^2}{2m} - \frac{1}{2} g\mu_B \sigma B \equiv \epsilon(\mathbf{p}, \sigma), \tag{5.12}$$

where the second term arises from the electromagnetic interaction [see Equations (5.5) and (5.8)]. Since $g = 2$ for the electron spin, we may simplify Equation (5.12) to

$$\epsilon = \frac{p^2}{2m} - \mu_B B\sigma. \tag{5.13}$$

This expression shows that the electron with up spin $(\sigma = +1)$ has a lower energy than the electron with down spin $(\sigma = -1)$. In other words, the spin degeneracy is removed in the presence of a magnetic field.

Let us now consider a collection of free electrons in equilibrium. At the absolute zero, the states with the lowest energies will be occupied by the electrons, the Fermi energy ϵ_F providing the upper limit. This situation is schematically shown in Figure 5.3, where the densities of states, $\mathcal{N}_+(\epsilon)$ and $\mathcal{N}_-(\epsilon)$, for electrons with up and down spins are drawn against the energy ϵ. The density of states for free electrons was discussed in Chapter 3. In the absence of the field, both $\mathcal{N}_+(\epsilon)$ and $\mathcal{N}_-(\epsilon)$ are the same and are given by

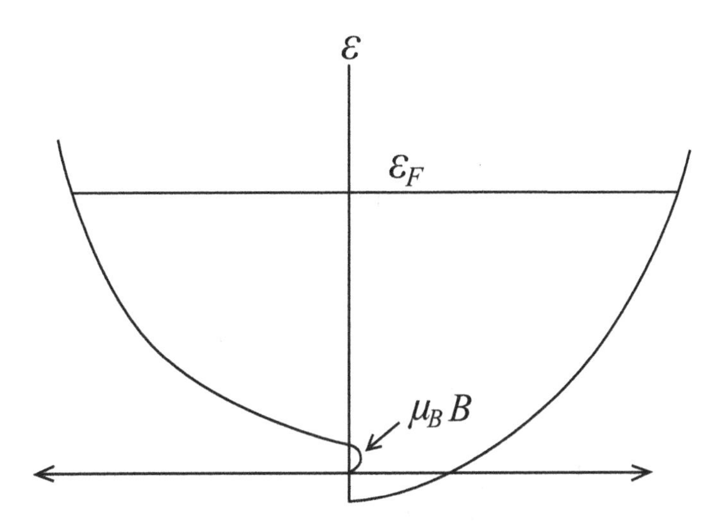

Figure 5.3: The density of states $\mathcal{N}_+$ and $\mathcal{N}_-$, for free electrons with up and down spins, are drawn against the energy ϵ, which is measured upwards.

one-half of expression (3.36):

$$\mathcal{N}_0(\epsilon) \equiv V \frac{m^{3/2}}{\sqrt{2}\,\pi^2\hbar^3}\epsilon^{1/2}. \tag{5.14}$$

Because of the magnetic energy $-\mu_B B$, the curve for the density of states, $\mathcal{N}_+(\epsilon)$, for electrons with up spins will be displaced downward by $\mu_B B$ compared with that for zero field and will be given by

$$\mathcal{N}_+(\epsilon) = V \frac{m^{3/2}}{\sqrt{2}\,\pi^2\hbar^3}(\epsilon + \mu_B B)^{1/2} = \mathcal{N}_0(\epsilon + \mu_B B), \quad \epsilon \geq -\mu_B B. \tag{5.15}$$

Similarly the curve for the density of states, $\mathcal{N}_-(\epsilon)$, for electrons with down spins is displaced upward by $\mu_B B$:

$$\mathcal{N}_-(\epsilon) = V \frac{m^{3/2}}{\sqrt{2}\,\pi^2\hbar^3}(\epsilon - \mu_B B)^{1/2} = \mathcal{N}_0(\epsilon - \mu_B B), \quad \epsilon \geq -\mu_B B. \tag{5.16}$$

From Figure 5.3, the numbers $N_\pm$ of the electrons with up and down spins are given by

$$N_+ = \int_{-\mu_B B}^{\epsilon_F} d\epsilon\, \mathcal{N}_+(\epsilon) = \int_0^{\epsilon_F + \mu_B B} dx\, \mathcal{N}_0(x) \qquad (x = \epsilon + \mu_B B),$$

$$N_- = \int_{\mu_B B}^{\epsilon_F} d\epsilon\, \mathcal{N}_-(\epsilon) = \int_0^{\epsilon_F - \mu_B B} dx\, \mathcal{N}_0(x) \qquad (x = \epsilon - \mu_B B)\,. \tag{5.17}$$

The difference $N_+ - N_-$ generates a finite magnetic moment for the system. Each electron with up spin contributes μ_B, and each electron with down spin contributes $-\mu_B$. Therefore, the total magnetic moment is $N_+\mu_B - N_-\mu_B$. Dividing this by volume V, we obtain, for the magnetization,

$$\begin{aligned} I &= \frac{\mu_B}{V}[N_+ - N_-] \\ &= \frac{\mu_B}{V}\left[\int_0^{\epsilon_F+\mu_B B} dx\, \mathcal{N}_0(x) - \int_0^{\epsilon_F-\mu_B B} dx\, \mathcal{N}_0(x)\right] \\ &\cong \frac{2\mu_B^2 B}{V}\mathcal{N}_0(\epsilon_F), \end{aligned} \tag{5.18}$$

where we retained the term proportional to B only. Using Equation (5.14), we can re-express this as follows:

$$I = \frac{\sqrt{2}\,\mu_B^2 m^{3/2}}{\pi^2\hbar^3}\epsilon_F^{1/2} B > 0. \tag{5.19}$$

The last expression shows that the magnetization is positive, and is proportional to the field B. That is, the system is paramagnetic. The *susceptibility* χ defined through the relation

$$I = \chi B \tag{5.20}$$

is given by

$$\chi = \frac{\sqrt{2}\,\mu_B^2 m^{3/2}}{\pi^2\hbar^3}\,\epsilon_F^{1/2}. \tag{5.21}$$

By using the relation

$$n = \frac{2}{3}\frac{\sqrt{2}\,m^{3/2}}{\pi^2\hbar^3}\,\epsilon_F^{3/2}, \tag{5.22}$$

we can rewrite Equation (5.21) as

$$\boxed{\chi_P = \frac{3}{2}\frac{\mu_B^2 n}{\epsilon_F}.} \tag{5.23}$$

This result was first obtained by Pauli [1] and is often referred to as the *Pauli paramagnetism*.

We note that the Pauli paramagnetism is weaker than the paramagnetism of isolated atoms approximately by the factor $k_B T/\epsilon_F$ (if this factor is small).

5.3 Landau Diamagnetism

The electron always circulates around the magnetic flux so as to reduce the magnetic field. This is called the *motional diamagnetism*. If we calculate this effect classically by considering the system confined to a closed volume, we obtain zero magnetic moments. This is known as *van Leeuwen's theorem*. We first demonstrate this theorem.

Let us take a system of free electrons confined in a volume V. The partition function per electron is

$$Z(B) = \frac{1}{(2\pi\hbar)^3} \int_V d^3r \int d^3p \exp\left[-\frac{|\mathbf{p}+e\mathbf{A}|^2}{2mk_BT}\right]. \tag{5.24}$$

We introduce kinetic momentum $\mathbf{\Pi}$:

$$m\dot{\mathbf{r}} \equiv m\mathbf{v} = \mathbf{p} + e\mathbf{A} \equiv \mathbf{\Pi} = (\Pi_x, \Pi_y, \Pi_z). \tag{5.25}$$

After simple calculations [see Equation (4.44)], we obtain

$$dx\,dy\,dz\,dp_x\,dp_y\,dp_z = dx\,dy\,dz\,d\Pi_x\,d\Pi_y\,d\Pi_z. \tag{5.26}$$

Using the last three equations, we can show that $Z(B)$ is equal to the electron partition function with no field;

$$Z(B) = Z(B=0). \tag{5.27}$$

This result depends on the container wall. In Figure 5.4 we show that counterclockwise currents due to the completed electron circulations are canceled out by the clockwise currents due to the incomplete circulations near the wall. Hence there are no magnetic moments.

In 1930, L.D. Landau [2] showed that the quantum treatment of the electron circulation yields a diamagnetic moment. We shall show this below.

Electrons obey the Fermi–Dirac statistics. Considering a system of free electrons, we define the *free energy* F as (Problems 5.3.1 and 5.3.2)

$$F = N\mu - 2k_BT \sum_i \ln\left[1 + e^{(\mu - E_i)/(k_BT)}\right], \tag{5.28}$$

where the factor 2 arises from the spin degeneracy. The chemical potential μ is determined from the condition

$$\frac{\partial F}{\partial \mu} = 0. \tag{5.29}$$

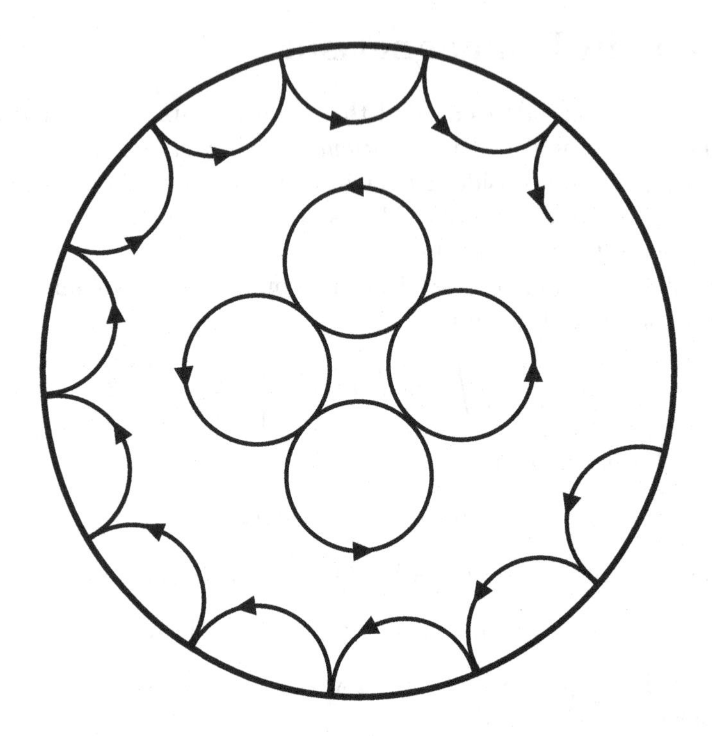

Figure 5.4: The diamagnetic currents due to the electron circulations inside are canceled out by the currents generated near the container wall.

The total magnetic moment M for the system can be found from

$$M = -\frac{\partial F}{\partial B}. \tag{5.30}$$

Equation (5.29) is equivalent to the usual condition (Problem 5.3.3) that the total number of the electrons, N, can be obtained in terms of the Fermi distribution function

$$f(E) \equiv \left[\exp\left(\frac{E-\mu}{k_B T} - 1\right) + 1\right]^{-1} \tag{5.31}$$

from

$$N = 2\sum_i f(E_i). \tag{5.32}$$

The Landau energy E_i is characterized by the Landau oscillator quantum number N_L and the z-component momentum p_z. The energy E becomes continuous in the bulk limit. Let us introduce the density of state $dW/dE = \mathcal{N}(E)$ such that

$$\mathcal{N}(E)dE = \text{number of states having an energy between } E \text{ and } E + dE. \tag{5.33}$$

We now write Equation (5.28) in the form

$$F = N\mu - 2k_B T \int_0^\infty dE \frac{dW}{dE} \ln\left[1 + e^{(\mu - E)/(k_B T)}\right]. \tag{5.34}$$

The *statistical weight* (number) W is the total number of states having energies less than

$$E = \left(N_L + \frac{1}{2}\right)\hbar\omega_c + \frac{p_z^2}{2m}. \tag{5.35}$$

Conversely, the allowed values of p_z are distributed over the range in which $|p_z|$ does not exceed

$$\left\{2m\left[E - \left(N_L + \frac{1}{2}\right)\hbar\omega_c\right]\right\}^{1/2}. \tag{5.36}$$

For a fixed pair (E, N_L) the increment in the weight, dW, is given by

$$\begin{aligned} dW &= eB\frac{L_1L_2}{(2\pi\hbar)} \int dp_z \frac{L_3}{(2\pi\hbar)} \\ &= VeB\frac{1}{(2\pi\hbar)^2}\, 2\left\{2m\left[E - \left(N_L + \frac{1}{2}\right)\hbar\omega_c\right]\right\}^{1/2}, \end{aligned} \tag{5.37}$$

where we used Equation (4.32); $V \equiv L_1L_2L_3$ is the volume of the container. After summing Equation (5.37) with respect to N_L, we obtain

$$W(E) = A\frac{(\hbar\omega_c)^{3/2}}{\sqrt{2\pi}}\, 2\sum_{N_L=0}^{\infty} \sqrt{\epsilon - (2N_L + 1)\pi}, \tag{5.38}$$

$$A \equiv V\frac{(2\pi m)^{3/2}}{(2\pi\hbar)^3}, \qquad \epsilon \equiv \frac{2\pi E}{\hbar\omega_c}.$$

We assume high Fermi degeneracy such that

$$\mu \simeq \epsilon_F \gg \hbar\omega_c \qquad \text{for a metal.} \tag{5.39}$$

The sum over N_L in Equation (5.38) converges slowly. We use *Poisson's summation formula* [3] and, after mathematical manipulations, obtain

$$W(E) = W_0 + W_L + W_{\text{osc}}, \tag{5.40}$$

$$W_0 = A\frac{4}{3\sqrt{\pi}}E^{3/2}, \tag{5.41}$$

$$W_L = -A\frac{1}{24\sqrt{\pi}}\frac{(\hbar\omega_c)^2}{E^{1/2}}, \tag{5.42}$$

$$W_{\text{osc}} = A\frac{(\hbar\omega_c)^{3/2}}{\sqrt{2}\pi^{3/2}}\sum_{\nu=1}^{\infty}\frac{(-1)^\nu}{\nu^{3/2}}\sin\left(\frac{2\pi\nu E}{\hbar\omega_c} - \frac{\pi}{4}\right). \tag{5.43}$$

The detailed steps leading to Equations (5.40) through (5.43) are given in Appendix A.

The term W_0, which is independent of B, gives the weight equal to that for a free-electron system with no field. The term W_L is negative (diamagnetic) and can generate a diamagnetic moment. We start with Equation (5.34), integrate by parts, and obtain

$$\begin{aligned} F &= N\mu - 2\int_0^\infty dE\, W(E)f(E) \\ &= N\mu + 2\int_0^\infty dE\frac{df}{dE}\int_0^E dE'\, W(E'). \end{aligned} \tag{5.44}$$

The $-df/dE$, which can be expressed as

$$-\frac{df}{dE} = \frac{1}{4k_BT}\,\text{sech}^2\left(\frac{E-\mu}{2k_BT}\right), \tag{5.45}$$

has a sharp peak near $E = \mu$ if $k_BT \ll \mu$, and

$$\int_0^\infty dE\frac{df}{dE} = -1. \tag{5.46}$$

For a smoothly changing integrand in the last member of Equations (5.44) $-df/dE$ can be regarded as a *Dirac delta-function*:

$$-\frac{df}{dE} = \delta(E - \mu). \tag{5.47}$$

Using this property and Equations (5.41) and (5.44), we obtain (Problem 5.3.4)

$$F_L = A\frac{1}{6\sqrt{\pi}}(\hbar\omega_c)^2 \epsilon_F^{1/2}. \tag{5.48}$$

Here we set $\mu = \epsilon_F$. This is justified since the corrections to $\mu(B, T)$ start with a B^4 term and with a T^2 term. Using Equations (5.48) and (5.30), we obtain

$$\boxed{\chi_L = V^{-1}\frac{\partial M_L}{\partial B} = -V^{-1}\frac{\partial^2 F_L}{\partial B^2} = -\frac{n\mu_B^2}{2\epsilon_F},} \tag{5.49}$$

where n is the free electron density.

Comparing this result with Equation (5.23), we observe that Landau diamagnetism is one third (1/3) of the Pauli paramagnetism in magnitude. But the calculations in this section are done with the assumption of the free electron model. If the effective mass (m^*) approximation is used, formula (5.49) is corrected by the factor $(m^*/m)^2$ as we see from Equation (5.48). Hence the diamagnetic susceptibility for a metal is

$$\chi_L^{\text{metal}} = (m/m^*)^2 \chi_L. \tag{5.50}$$

We note that the Landau susceptibility is spin-independent. For a metal having a small effective mass, the Landau susceptibility χ_L^{metal} can be greater in magnitude than the Pauli paramagnetic susceptibility χ_P. Then the total susceptibility expressed as

$$\chi = \chi_P + \chi_L^{\text{metal}} \tag{5.51}$$

can be negative (diamagnetic). This is observed for GaAs ($m^* = 0.07m$).

The oscillatory term W_{osc} in Equation (5.43) yields the de Haas-van Alphen oscillations, which will be discussed in Section 12.3.

Problem 5.3.1. The grand partition function Ξ is defined by

$$\Xi \equiv \mathrm{TR}\left\{\exp(\alpha N - \beta H)\right\} \equiv \sum_{N=0}^{\infty} e^{\alpha N} \mathrm{Tr}_{,N}\left\{\exp(\alpha N - \beta H_N)\right\},$$

where H_N is the Hamiltonian of the N-particle system and Tr stands for the trace (diagonal sum). The symbol TR means a grand ensemble trace. We assume that the internal energy E, the number density n, and the entropy S are given by

$$\begin{aligned} E &= \langle H \rangle \equiv \frac{\mathrm{TR}\left\{H \exp(\alpha N - \beta H)\right\}}{\Xi}, \\ n &= \frac{\langle N \rangle}{V} = \frac{\mathrm{TR}\left\{N \exp(\alpha N - \beta H)\right\}}{V\Xi}, \\ S &= k_B \left(\ln \Xi + \beta E - \alpha \langle N \rangle\right). \end{aligned}$$

Show that

$$F \equiv E - TS = \langle N \rangle \mu - k_B T \ln \Xi,$$

where $\mu \equiv k_B T \alpha$ is Gibbs free energy per particle: $G \equiv E - TS + PV = \mu \langle N \rangle$.

Problem 5.3.2.

a. Evaluate the grand partition function Ξ for a free-electron system characterized by

$$H = \sum_{j=1}^{N} \frac{p_j^2}{2m}.$$

b. Show that

$$n = V^{-1} \frac{\partial}{\partial \alpha} \ln \Xi = 2V^{-1} \sum_{\mathbf{p}} f(\epsilon_{\mathbf{p}}),$$

$$f(\epsilon) = \frac{1}{e^{\beta(\epsilon - \mu)} + 1}, \qquad \epsilon_p = \frac{p^2}{2m},$$

$$E = -\frac{\partial}{\partial \beta} \ln \Xi = 2 \sum_{\mathbf{p}} \epsilon_p f(\epsilon_p),$$

where 2 is the spin degeneracy factor.

Problem 5.3.3. Using the free energy F in Equation (5.28), obtain Equations (5.31) and (5.32).

Problem 5.3.4. Verify Equation (5.48).

Chapter 6

Boltzmann Equation Method

The Boltzmann equation is set up for a system of free electrons subject to the impurity scattering. By solving this equation, an exact expression for the conductivity is obtained.

6.1 The Boltzmann Equation

In the method of a Boltzmann equation the qualitative arguments in the simple kinetic theory will be formulated in more precise terms. In some simple cases, the description of the transport phenomena by this method is exact. In more complicated cases this method is an approximation, but it is widely used.

Let us consider the electron-impurity system, a system of free electrons with uniformly distributed impurities. We introduce a *momentum distribution function* $\phi(\mathbf{p}, t)$ defined such that $\phi(\mathbf{p}, t)d^3p$ gives the relative probability of finding an electron in the element d^3p at time t. This function will be normalized such that

$$\frac{2}{(2\pi\hbar)^3} \int d^3p \, \phi(\mathbf{p}, t) = \frac{N}{V} = n. \tag{6.1}$$

The *electric current density* $\mathbf{j}$ is given in terms of ϕ as follows:

$$\mathbf{j} = \frac{-2e}{(2\pi\hbar)^3 m} \int d^3p \, \mathbf{p} \, \phi(\mathbf{p}, t). \tag{6.2}$$

The function ϕ can be obtained by solving the Boltzmann equation, which may be set up in the following manner.

The change in the distribution function ϕ will be caused by the force acting on the electrons in the element d^3p *and* by the collision. We may write this change in the form

$$\frac{\partial \phi}{\partial t} = \left(\frac{d\phi}{dt}\right)_{\text{force}} + \left(\frac{d\phi}{dt}\right)_{\text{collision}} . \tag{6.3}$$

The *force term* $(d\phi/dt)_{\text{force}}$, caused by the force $-e\mathbf{E}$ acting on the electrons can be expressed by

$$\left(\frac{d\phi}{dt}\right)_{\text{force}} = -e\mathbf{E} \cdot \frac{\partial \phi}{\partial \mathbf{p}}. \tag{6.4}$$

If the density of impurities, n_I, is low and the interaction between electron and impurity has a short range, the electron will be scattered by one impurity at a time. We may then write the *collision term* in the following form:

$$\left(\frac{d\phi}{dt}\right)_{\text{collision}} = \int d\Omega \, \frac{p}{m} n_I I(p, \theta) [\phi(\mathbf{p}', t) - \phi(\mathbf{p}, \mathbf{t})], \tag{6.5}$$

where θ is the scattering angle, the angle between the initial and final momenta $\mathbf{p}$ and $\mathbf{p}'$, and $I(p, \theta)$ is the differential cross section. In fact, the rate of collision is given by (density of scatterers) $\times$ (speed) $\times$ (total cross section). If we apply this rule to the flux of particles with momentum $\mathbf{p}$, we can obtain the second integral of Equation (6.5), the integral with the minus sign. This integral corresponds to the *loss* of the flux due to the collision [see Figure 6.1 (a)]. The flux of particles with momentum $\mathbf{p}$ can *gain* by the inverse collision, which is shown in Figure 6.1 (b). The contribution of the *inverse collision* is represented by the first integral.

So far, we have neglected the fact that electrons are fermions and are therefore subject to the Pauli exclusion principle. We will now look at the effect of quantum statistics.

If the final momentum state $\mathbf{p}'$ was already occupied, then the scattering from the state $\mathbf{p}$ to the state $\mathbf{p}'$ should not have occurred. The probability of this scattering therefore should be reduced by the factor $1 - \phi(\mathbf{p}', t)$, which represents the probability that the final state $\mathbf{p}'$ is unoccupied. Consideration of the exclusion principle thus modifies the Boltzmann collision term given in Equation (6.5) to

$$\begin{aligned} \left(\frac{d\phi}{dt}\right)_{\text{collision}} = & \int d\Omega \, \frac{p}{m} n_I I(p, \theta) \left\{\phi(\mathbf{p}', t) \left[1 - \phi(\mathbf{p}, t)\right] \right. \\ & \left. - \phi(\mathbf{p}, t) \left[1 - \phi(\mathbf{p}', t)\right]\right\}. \end{aligned} \tag{6.6}$$

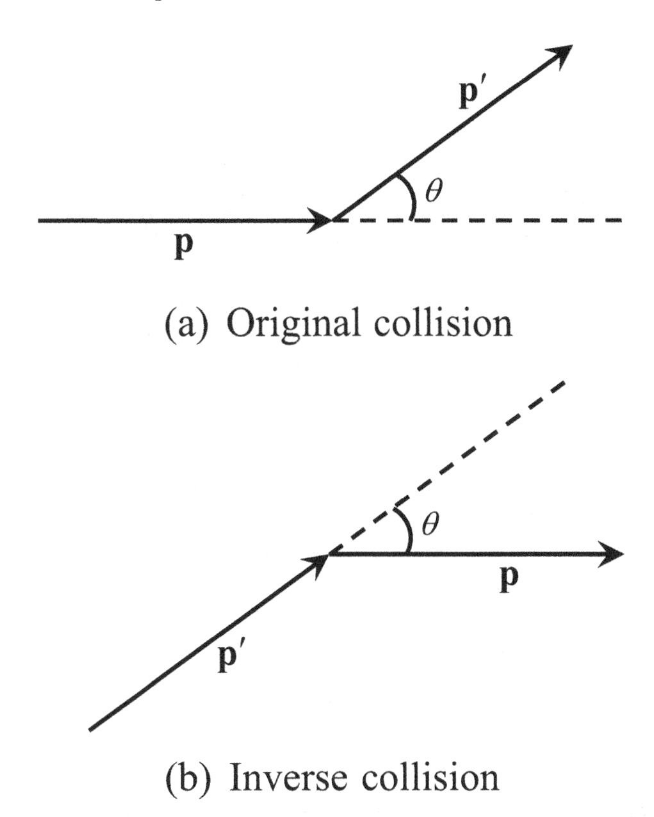

Figure 6.1: In (a), the electron suffers a change in momentum from $\mathbf{p}$ to $\mathbf{p}'$ after scattering. The inverse collision is shown in (b).

When Equation (6.6) is expanded, the two terms propotional to $\phi(\mathbf{p}', t)\phi(\mathbf{p}, t)$ in the curly brackets cancel each other out. We then have the same collision term as given by Equation (6.5).

Gathering the results from Equations (6.3) through (6.6), we obtain

$$\frac{\partial \phi(\mathbf{p}, t)}{\partial t} + (e\mathbf{E}) \cdot \frac{\partial \phi(\mathbf{p}, t)}{\partial \mathbf{p}} = \frac{n_I}{m} \int d\Omega \, p \, I(p, \theta)[\phi(\mathbf{p}', t) - \phi(\mathbf{p}, t)]. \tag{6.7}$$

This is the *Boltzmann equation* for the electron-impurity system. This equation is linear in ϕ, and much simpler than the Boltzmann equation for a dilute gas. In particular, we can solve Equation (6.7) by elementary methods and calculate the conductivity σ. We will do this in the next section. For simple forms of the scattering cross section, we can also solve Equation (6.7) as an initial-value problem (Problems 6.1.1 and 6.1.2).

As mentioned in the beginning of this section, the Boltzmann equation is very important, but it is an approximate equation. If the impurity density is high or if the range of the interaction is not short, then we must consider simultaneous scatterings by two or more impurities. If we include the effect of the coulomb interaction among electrons, which has hitherto been ignored, the collision term should be further modified. It is, however, very difficult to estimate appropriate corrections arising from these various effects.

Problem 6.1.1. Obtain the Boltzmann equation for an electron-impurity system in two dimensions, which may be deduced by inspection, from Equation (6.7). Assmue that all electrons have the same velocity at the initial time $t = 0$. Further, assuming no electric field ($E = 0$) and isotropic scattering ($I =$ constant), solve the Boltzmann equation.

Problem 6.1.2. Solve the Boltzmann equation (6.7) for three dimensions with the same condition as in Problem 6.1.1. Define the Boltzmann H-function by $H(t) \equiv (1/n) \int d^3p \; \phi(\mathbf{p}, t) \ln[\phi(\mathbf{p}, t)/n]$. Evaluate dH/dt. Plot $H(t)$ as a function of time.

6.2 The Current Relaxation Rate

Let us assume that a small constant electric field $\mathbf{E}$ is applied to the electron-impurity system and that a stationary homogeneous current is established. We take the positive x-axis along the field $\mathbf{E}$. In the stationary state, $\partial\phi/\partial t = 0$: the distribution function ϕ depends on momentum $\mathbf{p}$ only. From Equation (6.7) the Boltzmann equation for ϕ is then given by

$$-eE\frac{\partial \phi(\mathbf{p})}{\partial p_x} = \frac{n_I}{m} \int d\Omega \; pI[\phi(\mathbf{p}') - \phi(\mathbf{p})]. \tag{6.8}$$

We wish to solve this equation and calculate the conductivity σ.

In the absence of the field $\mathbf{E}$, the system, by assumption, is characterized by the equilibrium distribution function, that is, the Fermi distribution function for free electrons:

$$\phi_0(\mathbf{p}) = f(\epsilon_p) \equiv \frac{1}{e^{\beta(\epsilon_p - \mu)} + 1}. \tag{6.9}$$

With the small field E, the function ϕ deviates from ϕ_0. Let us regard ϕ as a function of E and expand it in powers of E:

$$\begin{aligned}\phi(\mathbf{p}) &= \phi_0 + \phi_1 + \cdots \\ &= f(\epsilon_p) + \phi_1(\mathbf{p}) + \cdots ,\end{aligned} \tag{6.10}$$

where the subscripts denote the orders in E. For the determination of the conductivity σ we need ϕ_1 only. Let us introduce Equation (6.10) in Equation (6.8), and compare terms of the same order in E.

In the zeroth order we have

$$\frac{n_I}{m}\int d\Omega\, pI[\phi_0(\mathbf{p}') - \phi_0(\mathbf{p})] = 0. \tag{6.11}$$

Since the energy is conserved in each scattering,

$$\epsilon_{p'} = \epsilon_p, \tag{6.12}$$

$\phi_0(\mathbf{p}) \equiv f(\epsilon_p)$ clearly satisfies Equation (6.11). In the first order in E, we have

$$\begin{aligned}\text{LHS} &= -eE\frac{\partial \phi_0}{\partial p_x} = -eE\frac{\partial f(\epsilon_p)}{\partial p_x} \\ &= -eE\frac{\partial \epsilon_p}{\partial p_x}\frac{df}{d\epsilon_p} = -eE\frac{p_x}{m}\frac{df}{d\epsilon_p}.\end{aligned}$$

Therefore, we obtain from Equation (6.8)

$$-eE\frac{p_x}{m}\frac{df}{d\epsilon_p} = \frac{n_I}{m}\int d\Omega\, pI[\phi_1(\mathbf{p}') - \phi_1(\mathbf{p})]. \tag{6.13}$$

For the moment, let us neglect the first term on the RHS. In this case, the RHS equals $-n_I m^{-1} p\phi_1(\mathbf{p}) \int d\Omega\, I$. Then, the function $\phi_1(\mathbf{p})$ is proportional to p_x. Let us now try a solution of the form

$$\phi_1(\mathbf{p}) = p_x \Phi(\epsilon_p), \tag{6.14}$$

where $\Phi(\epsilon_p)$ is a function of ϵ_p (no angular dependence). Substitution of Equation (6.14) into Equation (6.13) yields

$$-eEp_x \frac{df}{d\epsilon_p} = n_I \int d\Omega\, pI\Phi(\epsilon_p)(p'_x - p_x). \tag{6.15}$$

Let us look at the integral on the RHS. We introduce a new frame of reference with the polar axis (the z-axis) pointing along the fixed vector $\mathbf{p}$ as shown in Figure 6.2. The old positive x-axis, which is parallel to the electric field $\mathbf{E}$, can be specified by the angle (θ, ϕ). From the diagram we have

$$p_x = p\cos\theta. \tag{6.16}$$

The vector $\mathbf{p}'$ can be represented by (p, χ, ϕ_1). If we denote the angle between $\mathbf{p}'$ and $\mathbf{E}$ by ψ, we have

$$p'_x = p\cos\psi. \tag{6.17}$$

To express $\cos\psi$ in terms of the angles (θ, ϕ) and (χ, ϕ_1), we use the vector decomposition property and obtain

$$\begin{aligned}\cos\psi &= \frac{\mathbf{p}'}{p}\cdot\frac{\mathbf{E}}{E} \\ &= (\mathbf{i}\sin\chi\cos\phi_1 + \mathbf{j}\sin\chi\sin\phi_1 + \mathbf{k}\cos\chi) \\ &\quad\cdot(\mathbf{i}\sin\theta\cos\phi + \mathbf{j}\sin\theta\sin\phi + \mathbf{k}\cos\theta) \\ &= \sin\chi\sin\theta[\cos\phi_1\cos\phi + \sin\phi_1\sin\phi] + \cos\chi\cos\theta\end{aligned}$$

or

$$\cos\psi = \sin\chi\sin\theta\cos(\phi - \phi_1) + \cos\chi\cos\theta. \tag{6.18}$$

Let us now consider the first integral in Equation (6.15):

$$\begin{aligned}A &\equiv n_I \int d\Omega\, pI(p,\chi)\Phi(\epsilon_p)(p\cos\psi) \\ &= n_I \int_0^{2\pi} d\phi_1 \int_0^{\pi} d\chi\; \sin\chi\; p^2 I(p,\chi)\Phi(\epsilon_p) \\ &\quad\times[\sin\chi\sin\theta\cos(\phi - \phi_1) + \cos\chi\cos\theta].\end{aligned} \tag{6.19}$$

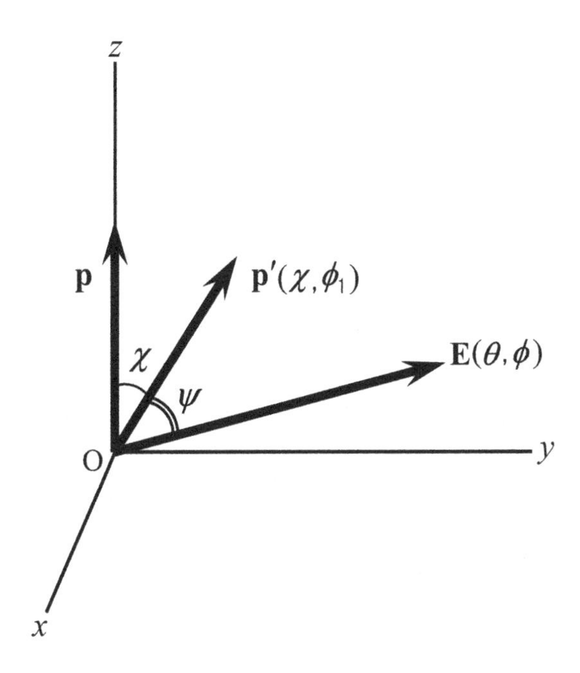

Figure 6.2: A new frame of reference in which the positive z-axis points in the direction of the fixed vector $\mathbf{p}$. In this frame, the direction of the electric field $\mathbf{E}$ is specified by (θ, ϕ) and that of the momentum $\mathbf{p}'$ by (χ, ϕ_1).

Since

$$\int_0^{2\pi} d\phi_1 \cos(\phi - \phi_1) = 0, \tag{6.20}$$

the first integral can be dropped. We then obtain

$$\begin{aligned} A &= n_I \int_0^{2\pi} d\phi_1 \int_0^{\pi} d\chi \sin\chi \; pI(p,\chi) \cos\chi \; [p\cos\theta \Phi(\epsilon_p)] \\ &= \phi_1(\mathbf{p}) n_I \int d\Omega \; pI(p,\chi)\cos\chi. \end{aligned} \tag{6.21}$$

The second integral in Equation (6.15), therefore, is proportional to $\phi_1(\mathbf{p})$. We thus obtain the solution:

$$\phi_1(\mathbf{p}) = eE \frac{p_x}{m} \frac{df}{d\epsilon_p} \frac{1}{\Gamma(p)}, \tag{6.22}$$

$$\Gamma(p) \equiv \frac{n_I}{m} \int d\Omega\, pI(p,\chi)[1 - \cos\chi] > 0. \tag{6.23}$$

The Γ here is positive and depends only on $p \equiv |p|$ (or equivalently on the energy ϵ_p); it has the dimension of frequency and is called the *energy-dependent current relaxation rate* or simply the *relaxation rate*. Its inverse is called the *relaxation time*.

The electric current density j_x can be calculated from

$$j_x = \frac{-2e}{(2\pi\hbar)^3 m} \int d^3p\, p_x \phi(\mathbf{p}). \tag{6.24}$$

We introduce $\phi = \phi_0 + \phi_1 + \cdots$ in this expression. The first term gives a vanishing contribution. (No current in equilibrium.) The second term yields, using Equation (6.22),

$$j_x = -\frac{2e^2 E}{(2\pi\hbar)^3 m^2} \int d^3p\, \frac{p_x^2}{\Gamma(p)} \frac{df}{d\epsilon_p}. \tag{6.25}$$

Comparing this with Ohm's law, $j_x = \sigma E$, we obtain the following expression for the conductivity:

$$\sigma = \frac{2e^2}{(2\pi\hbar)^3 m^2} \int d^3p\, \frac{p_x^2}{\Gamma(p)} \left(-\frac{df}{d\epsilon_p}\right) = \frac{4}{3} \frac{e^2}{m(2\pi\hbar)^3} \int d^3p\, \frac{1}{\Gamma(p)}\, \epsilon\, \frac{df}{d\epsilon}, \tag{6.26}$$

where we used $\epsilon = (2m)^{-1}(p_x^2 + p_y^2 + p_z^2)$. A few applications of this formula will be discussed later.

Problem 6.2.1. For a classical hard-sphere interaction, the scattering cross section I is given by $(1/2)a^2$, where a is the radius of the sphere. Evaluate the relaxation rate $\Gamma(p)$ given by Equation (6.23). Using this result, calculate the conductivity σ through Equation (6.26). Verify that the conductivity calculated is temperatureindependent.

Problem 6.2.2. Formula (6.26) was obtained with the assumption that the equilibrium distribution function in the absence of the field is given by the Fermi distribution function, Equation (6.9).

a. Verify that the same formula applies when we assume that the equilibrium distribution function is given by the Boltzmann distribution function.
b. Show that the conductivity calculated by this formula does not depend on the temperature.

Part II

Bloch Electron Dynamics

We discuss the Bloch electron dynamics in Part II, Chapters 7 through 10. To properly develop the theory of conduction electron dynamics, a deeper understanding of the properties of normal metals beyond the free-electron model is required. Based on the Bloch's theorem, the Fermi liquid model is derived. At 0 K, the normal metal is shown to have a sharp Fermi surface, which is experimentally supported by the fact that the heat capacity is linear in temperature at the lowest temperatures. "Electrons" ("holes"), which by definition circulate counterclockwise (clockwise) viewed from the tip of the magnetic field vector, play major roles in metal and semiconductor physics. These conduction electrons ("electrons," "holes") are generated, depending on the curvature of the Fermi surface. From this viewpoint a *new* theory of the Bloch electron dynamics is developed.

Chapter 7

Bloch Theorem

The Bloch theorem plays a central role in conduction electron dynamics. The theorem is derived and discussed in this chapter.

7.1 The Bloch Theorem

Let us consider a periodic potential $V(x)$ in one dimension [see Figure 7.1] that satisfies

$$V(x+na) = V(x), \qquad -\infty < x < \infty, \tag{7.1}$$

where a is the lattice constant and n is an integer. The Schrödinger energy-eigenvalue equation for an electron is

$$\left[-\frac{\hbar^2}{2m}\frac{d^2}{dx^2} + V(x)\right]\psi_E(x) = E\psi_E(x). \tag{7.2}$$

Clearly the wave function $\psi_E(x+na)$ also satisfies the same equation. Therefore, $\psi_E(x+na)$ is likely to be different from $\psi_E(x)$ only by an x-independent phase:

$$\boxed{\psi_E(x+na) = e^{ikna}\psi_E(x),} \tag{7.3}$$

where k is a real number, see below. Equation (7.3) represents a form of the *Bloch theorem* [1]. It generates far-reaching consequences in the theory of conduction electrons. Let us prove Equation (7.3). Since $\psi(x)$ and $\psi(x+na)$

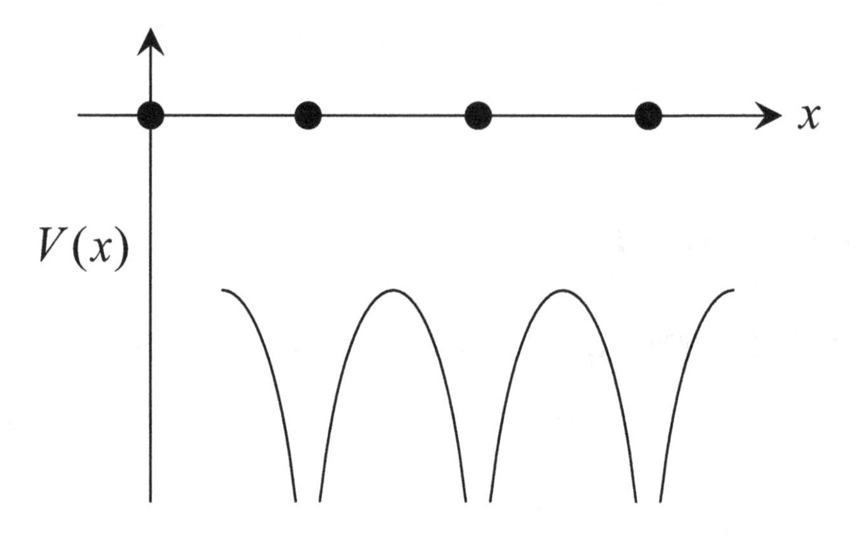

Figure 7.1: A periodic potential $V(x)$ in one dimension.

satisfy the same equation, they are linearly dependent:

$$\psi(x + na) = c(na)\psi(x). \tag{7.4}$$

Using Equation (7.4) twice, we obtain

$$\psi(x + na + ma) = c(na)\psi(x + ma) = c(na)c(ma)\psi(x) = c(na + ma)\psi(x).$$

Since the wave function $\psi(x)$ does not vanish in general, we obtain

$$c(na + ma) = c(na)c(ma) \quad \text{or} \quad c(x + y) = c(x)c(y). \tag{7.5}$$

Solving this functional equation, we obtain (Problem 7.1.1)

$$c(y) = \exp(\lambda y), \tag{7.6}$$

where λ is a constant. Because the wave function ψ in Equation (7.4) must be finite for all ranges, constant λ must be a pure imaginary number:

$$\lambda = ik, \tag{7.7}$$

where k is real. Combining Equations (7.4), (7.6), and (7.7), we obtain Equation (7.3). QED.

Let us discuss a few physical properties of the Bloch wave function ψ. By taking the absolute square of Equation (7.3), we obtain

$$|\psi(x+na)|^2 = |\psi(x)|^2. \tag{7.8}$$

The following three main properties are observed.

A. The probability distribution function $P(x) \equiv |\psi(x)|^2$ is *lattice-periodic*:

$$P(x) \equiv |\psi(x)|^2 = P(x+na)\,, \quad \text{for any } n. \tag{7.9}$$

B. The exponential function of a complex number e^{iy} (y real) is periodic: $e^{i(y+2\pi m)} = e^{iy}$, where m is an integer. We may choose the real number k in Equation (7.3), called the k-number (2π times the wave number), to have a *fundamental range*:

$$-\frac{\pi}{a} < k < \frac{\pi}{a}; \tag{7.10}$$

the two end points are called the *Brillouin boundary* (points).

C. In general, there are a number of energy gaps (forbidden regions of energy) in which no solutions of Equation (7.2) exist (see Figure 7.4 and Section 7.2). The energy eigenvalues E are characterized by the k-number and the *zone number* (or *band index*) j, which enumerates the *energy bands*:

$$E = E_j(k). \tag{7.11}$$

This property C is not obvious, and it will be illustrated by examples in Section 7.2.

To further explore the nature of the Bloch wave function ψ, let us write

$$\boxed{\psi_E(x) = e^{ikx}u_{j,k}(x),} \tag{7.12}$$

and substitute it into Equation (7.3). If the function $u_{j,k}(x)$ is lattice periodic,

$$u_{j,k}(x+na) = u_{j,k}(x), \tag{7.13}$$

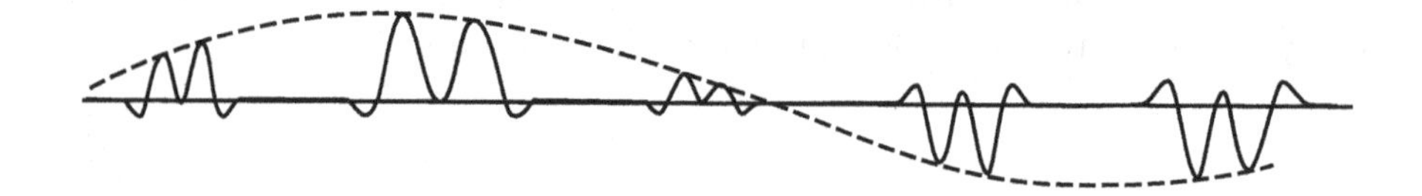

Figure 7.2: Variation of the real (or imaginary) part of the wave function $\psi_E(x)$.

then Equation (7.3) is satisfied (Problem 7.1.2). Equation (7.12) represents a second form of the Bloch theorem. The Bloch wave function $\psi(x) = e^{ikx}u_{j,k}(x)$ has great similarity with the free-particle wave function

$$\psi_{\text{free}}(x) = c\exp(ikx), \tag{7.14}$$

where c is a constant. The connection may be illustrated as shown in Figure 7.2. For the free particle, the k-number can range from $-\infty$ to ∞, and the energy is

$$E_{\text{free}} = \frac{p^2}{2m} \equiv \frac{\hbar^2 k^2}{2m} \tag{7.15}$$

with no gaps. These features are different from the properties B and C.

An important similarity arises when we write the time-dependent wave function $\psi(x,t)$ in the running wave form:

$$\psi_E(x,t) = e^{i(kx-\omega t)}U(x), \tag{7.16}$$

where the frequency ω is defined by

$$\omega = \begin{cases} \hbar^{-1}E_j(k) & \text{for the Bloch electron} \\ \hbar^{-1}E_{\text{free}} & \text{for the free electron,} \end{cases} \tag{7.17}$$

and the amplitude $U(x)$ is defined by

$$U(x) = \begin{cases} u_{j,k}(x) & \text{for the Bloch electron} \\ c & \text{for the free electron.} \end{cases} \tag{7.18}$$

Equation (7.16) shows that the Bloch wave function $\psi_E(x)$ represents a *running wave* characterized by k-number k, angular frequency ω, and *wave train* $u_{j,k}$.

The *group velocity* v of the Bloch wave packet is given by

$$v \equiv \frac{\partial \omega}{\partial k} \equiv \hbar^{-1} \frac{\partial E}{\partial k}. \tag{7.19}$$

By applying the (quantum) principle of wave-particle duality, we say that the Bloch electron moves with the *dispersion* (energy-momentum) *relation*:

$$E = \epsilon_j(\hbar k) \equiv \epsilon_j(p). \tag{7.20}$$

The velocity v is given by Equation (7.19). This gives a picture of great familiarity. Before fully developing this picture, we generalize our theory to the three-dimensional case.

Let us consider an infinite orthorhombic (ORC) lattice of lattice constants $(a,\ b,\ c)$. We choose a Cartesian frame of coordinates $(x,\ y,\ z)$ along the lattice axes. The potential $V(x,\ y,\ z) = V(\mathbf{r})$ is *lattice-periodic*:

$$V(\mathbf{r} + \mathbf{R}) = V(\mathbf{r}), \tag{7.21}$$

$$\mathbf{R} \equiv n_1 a\mathbf{i} + n_2 b\mathbf{j} + n_3 c\mathbf{k} \quad (n_j \text{ are integers}). \tag{7.22}$$

Vector $\mathbf{R}$ is called a *Bravais lattice vector*. The Schrödinger equation is

$$\left[-\frac{\hbar^2}{2m} \nabla^2 + V(\mathbf{r}) \right] \psi_E(\mathbf{r}) = E \psi_E(\mathbf{r}). \tag{7.23}$$

The Bloch wave function $\psi_E(\mathbf{r})$ satisfies

$$\boxed{\psi_E(\mathbf{r} + \mathbf{R}) = e^{i\mathbf{k}\cdot\mathbf{R}}\, \psi_E(\mathbf{r}),} \tag{7.24}$$

where $\mathbf{k} = (k_x,\ k_y,\ k_z)$ are called *k-vectors*.

The three principal properties of the Bloch wave function are as follows:

A. The probability distribution $P(\mathbf{r})$ is lattice-periodic:

$$P(\mathbf{r}) \equiv |\psi(\mathbf{r})|^2 = P(\mathbf{r} + \mathbf{R}). \tag{7.25}$$

B. The k-vector $\mathbf{k} = (k_x, k_y, k_z)$ in Equation (7.24) has the fundamental range

$$-\frac{\pi}{a} < k_x < \frac{\pi}{a}, \quad -\frac{\pi}{b} < k_y < \frac{\pi}{b}, \quad -\frac{\pi}{c} < k_z < \frac{\pi}{c}; \tag{7.26}$$

the endpoints, which form a rectangular box, are called the *Brillouin boundary*.

C. The energy eigenvalues E have energy gaps, and the allowed energies E can be characterized by the zone number j and the k-vectors:

$$\boxed{E = \epsilon_j(\hbar\mathbf{k}) \equiv \epsilon_j(\mathbf{p}),} \tag{7.27}$$

Equations (7.24) to (7.27) are straightforward extensions of Equations (7.3), (7,8), (7.10), and (7.11). Using Equation (7.24), we can express the Bloch wave function ψ in the form

$$\boxed{\psi_E(\mathbf{r}) \equiv \psi_{j,\mathbf{k}}(\mathbf{r}) = e^{i\mathbf{k}\cdot\mathbf{r}} u_{j,k}(\mathbf{r}), \quad u_{j,k}(\mathbf{r}+\mathbf{R}) = u_{j,k}(\mathbf{r}).} \tag{7.28}$$

Equations (7.27) and (7.28) indicate that the Bloch wave function $\psi_E(\mathbf{r})$, associated with quantum numbers $(j, \mathbf{k})$, is a plane wave characterized by k-vector $\mathbf{k}$, angular frequency $\omega \equiv \hbar^{-1}E_j(\mathbf{k})$, and wave train $u_{j,k}(\mathbf{r})$.

Problem 7.1.1. Solve the following functional equations:

$$\text{a.} \quad f(x+y) = f(x) + f(y), \tag{A}$$
$$\text{b.} \quad f(x+y) = f(x)f(y). \tag{B}$$

Hints: Differentiate Equation (A) with respect to x, and convert the result into an ordinary differential equation. Take the logarithm of Equation (B), and use (a). [Answer: (a) $f(x) = cx$, (b) $f(x) = e^{\lambda x}$.]

Problem 7.1.2. (a) Show that a wave equation of the Bloch form Equation (7.12) satisfies Equation (7.3). (b) Assuming Equation (7.3), derive Equation (7.12).

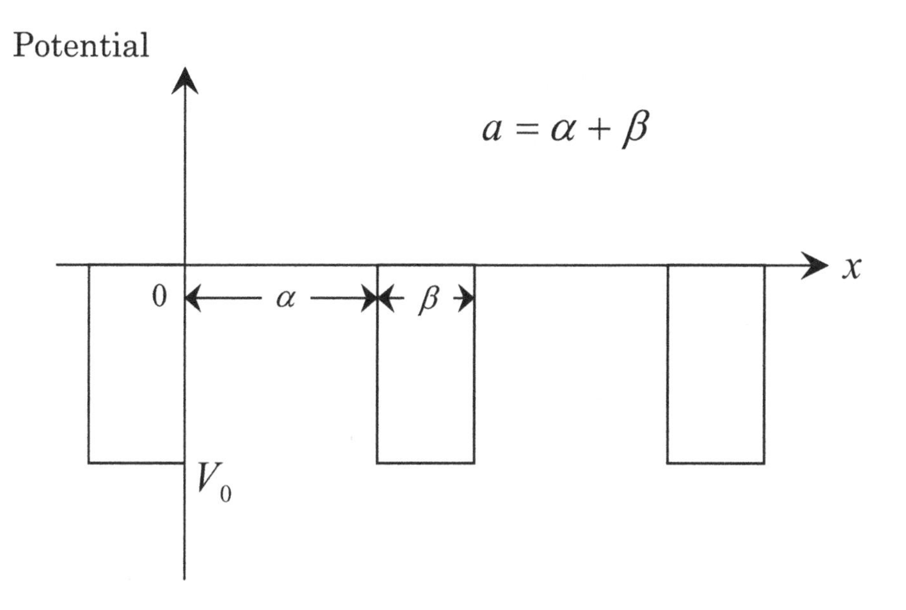

Figure 7.3: A Kronig–Penney potential.

7.2 The Kronig–Penney Model

The Bloch energy-eigenvalues in general have bands and gaps. We show this by taking the *Kronig–Penney* (K-P) *model* [2]. Let us consider a periodic square-well potential $V(x)$ with depth V_0 (< 0) and well width $\beta \equiv a - \alpha$ as shown in Figure 7.3:

$$V(x) = \begin{cases} V_0 & \text{if} \quad na - \beta < x < na \\ 0 & \text{if} \quad na < x < na + \alpha, \quad \alpha + \beta = a. \end{cases} \tag{7.29}$$

The Schrödinger energy eigenvalue equation for an electron can be written as Equation (7.2). Since this is a linear homogeneous differential equation with constant coefficients, the wave function $\psi(x)$ should have the form

$$\psi(x) = c\, e^{\gamma x}, \qquad (c, \gamma \text{ are constants}). \tag{7.30}$$

According to the Bloch theorem in Equation (7.12), this function $\psi(x)$ can be written as

$$\psi_k(x) = e^{ikx}\, u_k(x), \tag{7.31}$$

$$u_k(x+a) = u_k(x). \tag{7.32}$$

The condition that the function $\psi(x)$ be *continuous* and *analytic* at the well boundary yields the following relationships (Problem 7.2.1):

$$\cos ka = \cosh K\alpha \cos \mu\beta - \frac{\mu^2 - K^2}{2K\mu} \sinh K\alpha \sin \mu\beta \equiv f(E), \tag{7.33}$$

$$E \equiv -\frac{\hbar^2}{2m}K^2 \equiv V_0 + \frac{\hbar^2}{2m}\mu^2 \quad (V_0 < 0). \tag{7.34}$$

By solving Equation (7.33) with Equation (7.34), we can obtain the eigenvalue E as a function of k. The *band edges* are obtained from

$$f(E) = \pm 1, \tag{7.35}$$

which corresponds to the limits of $\cos ka$. Numerical studies of Equation (7.33) indicate [3] that (1) there are, in general, a number of negative- and positive-energy bands; (2) at each band edge, an *effective mass* m^* can be defined, whose value can be positive or negative and whose absolute value can be greater or less than the electron mass m; and (3) the effective mass is positive at the lower edge of each band, and it is negative at the upper edge. A typical dispersion relation for the model, showing energy bands and energy gaps, are shown in Figure 7.4. At the lowest band edge ϵ_0 we have

$$f(\epsilon_0) = 1. \tag{7.36}$$

Near this edge the dispersion (energy-k) relation calculated from Equation (7.33) is (Problem 7.2.2)

$$\epsilon = \epsilon_0 + \frac{\hbar^2}{2m^*}k^2 \quad (\epsilon_0 < 0), \tag{7.37}$$

$$m^* \equiv -\hbar^2 a^{-2} f'(\epsilon_0). \tag{7.38}$$

This one-dimensional K-P model can be used to study a simple three-dimensional model. Let us take an ORC lattice of unit lengths (a_1, a_2, a_3), with each lattice point representing a short-range attractive potential center (ion). The Schrödinger equation (7.23) for this system is hard to solve.

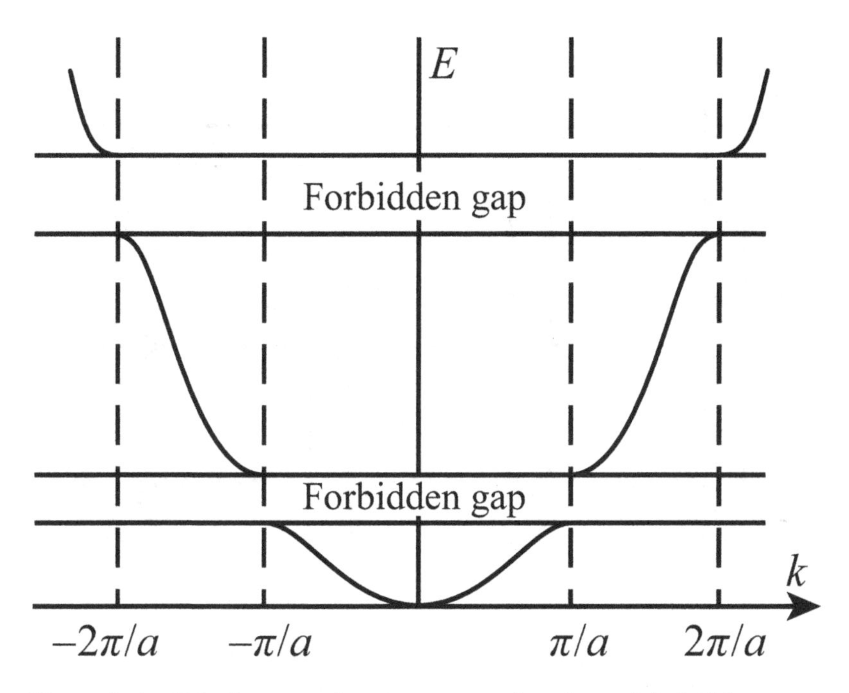

Figure 7.4: E-k diagram showing energy bands and forbidden gaps.

Let us now construct a model potential V_s defined by

$$V_s(x, y, z) = V_1(x) + V_2(y) + V_3(z), \tag{7.39}$$

$$V_j(u) = \begin{cases} V_0(< 0) & \text{if } na_j - \beta < u < na_j \\ 0 & \text{otherwise.} \end{cases} \tag{7.40}$$

Here the n are integers. A similar two-dimensional model is shown in Figure 7.5. The domains in which $V_s \neq 0$ are parallel plates of thickness β ($< a_j$) separated by a_j in the direction x_j, $(x_1, x_2, x_3) \equiv (x, y, z)$. The intersection of any two plates are straight beams of cross section β^2, where the potential V_s has the value $2V_0$. The intersections of three plates, where the potential V_s has the value $3V_0$, are cubes of side length β. The set of these cubes form an ORC lattice, a configuration similar to that of the commercially available molecular lattice model made up of balls and sticks. Note: Each square-well potential V_j has three parameters (V_0, β, a_j), and this model represents the

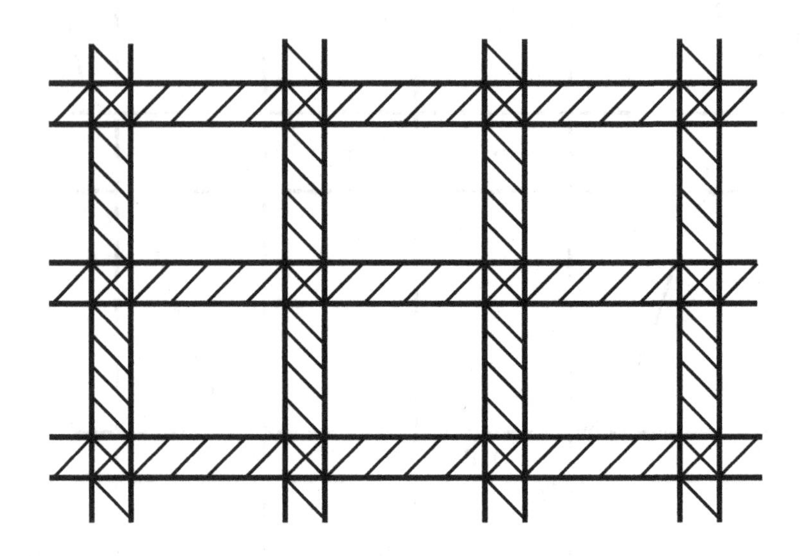

Figure 7.5: A 2D model potential. Each singly shaded strip has a potential energy (depth) V_0. Each cross-shaded square has a potential energy $2V_0$.

true potential fairly well [4]. The Schrödinger equation for the 3D model Hamiltonian

$$\begin{aligned} H_s &\equiv \frac{p_x^2 + p_y^2 + p_z^2}{2m} + V_1(x) + V_2(y) + V_3(z) \\ &= \frac{p_x^2}{2m} + V_1(x) + \frac{p_y^2}{2m} + V_2(y) + \frac{p_z^2}{2m} + V_3(z) \end{aligned} \tag{7.41}$$

can now be reduced to three one-dimensional K-P equations. We can then write an expression for the energy of our model system near the lowest band edge as

$$E = \frac{\hbar^2}{2m_1}k_x^2 + \frac{\hbar^2}{2m_2}k_y^2 + \frac{\hbar^2}{2m_3}k_z^2 + \text{constant}, \tag{7.42}$$

where $\{m_j\}$ are effective masses defined by Equation (7.38) with $a = a_j$.

Equation (7.42) is identical to what is intuitively expected of the energy-k relation for electrons in the ORC lattice. It is stressed that we *derived* it from first principles assuming a three-dimensional model Hamiltonian H_s. Our study demonstrates qualitatively how electron energy bands and gaps are generated from the Schrödinger equation for a Bloch electron moving in a three-dimensional lattice.

Problem 7.2.1. Derive Equation (7.33).

Problem 7.2.2. Derive Equation (7.38).

Chapter 8

The Fermi Liquid Model

The normal metal has a sharp Fermi surface at 0 K in spite of the fact that the electron motion is correlated due to the coulomb interaction. The independent electron model with a sharp Fermi surface at 0 K is called the Fermi liquid model. We derive this model in the self-consistent mean field approximation in this chapter.

8.1 The Self-consistent Field Approximation

The most important guide in chemistry is *Mendeleev's periodic table*. The binding (ionization) energy of atoms exhibits a periodic behavior.

First, we consider a hydrogen atom H, which is composed of an electron and a proton. The Hamiltonian of the system is the kinetic energies of the electron and the proton plus the coulomb attraction energy between the two particles. We introduce the center of mass (CM) and the relative coordinates and separate the Hamiltonian in two parts: the Hamiltonians describing the CM motion and the relative motion.

We write down the Schrödinger eigenvalue equation for the relative motion. The bound eigenstates are characterized by three quantum numbers $(n,\ l,\ m)$. They are respectively called the *principal*, *angular momentum*, and *magnetic* (or *azimuthal*) quantum numbers. They are all integers, but they are subject to the following restrictions. First, the principal quantum numbers n are positive integers $1,\ 2,\ \ldots$. For a fixed n, angular momentum quantum numbers l are $n-1,\ n-2,\ \ldots$, terminating with zero. For a fixed l, magnetic quantum numbers m are $-l,\ -l+1,\ \ldots$, terminating with $+l$.

The first few $(n,\ l,\ m)$ are $n = 1$, $l = 0$, $m = 0$; $n = 2$, $l = 0$, $m = 0$; $n = 2$, $l = 1$, $m = -1, 0, 1$; $n = 3$, $l = 0$, $m = 0$; $n = 3$, $l = 1$, $m = -1, 0, 1$; and $n = 3$, $l = 2$, $m = -2, -1, 0, 1, 2$. The energies depend neither on the angular momentum quantum number l nor on the magnetic quantum number m. The ground state of H is represented by $n = 1$, $l = 0$, and $m = 0$. The relative motion part, which contains the bound (negative-energy) states, is important in chemistry. Thus, we say that the ground state is represented by the atomic orbitals 1s with s meaning $l = 0$.

Second, consider a helium (He) atom, which is composed of two electrons and an ion (nucleus) He^{2+}. Let us focus on one of the electrons. This electron moves in the medium consisting of the ion He^{2+} and an electron, with the net charge $+e$. The medium's charge distribution is centered at the nucleus. Thus the electron moves in a hydrogenlike environment. This happens whichever electron we choose. We therefore find that the electron moves in a hydrogenlike self-consistent (mean) field.

An electron has a half spin whose state is represented by *two* values of the z-component along the magnetic field: $s_z = \pm(1/2)\hbar \equiv (1/2)\hbar\sigma$, $\sigma = \pm 1$. In the absence of the field the electron's energy is degenerate.

In summary, the state of the electron for a hydrogen is represented by the four quantum numbers $(n,\ l,\ m,\ \sigma)$. In the absence of the magnetic field, the energy can be represented by the two quantum numbers $(n,\ l)$. According to the Pauli exclusion principle no two electrons can occupy the same state. Hydrogenlike states are filled up one by one by electrons, starting with the lowest energy state 1s. For He the ground-state *atomic orbitals* are represented by $(1s)^2$. The index 2 comes from the spin degeneracy. Neon (^{10}Ne) has 10 electrons, and the atomic orbitals are represented by $1s^2\ 2s^2\ 2p^6$. The orbital 2p means that $n = 2$ and $l = 1$. The degeneracy in m is given by $2l+1 = 3$. The spin degeneracy doubles 3 to 6, explaining the index 6 on 2p. The chemical property of an atom such as the ionicity is determined by the outermost electron, which may be taken away by an outside means. He and Ne are inert atoms since the outermost shells $(1s^2)$ and $(2s^2\ 2p^6)$ are filled up and the ionization energies are high. There are eight electrons in the $n = 2$ states $(2s^2\ 2p^6)$, which explains the *eight columns* in the periodic table. Lithium ^3Li $(1s^2\ 2s)$ has a lone electron in the 2s orbital, can easily lose it (hence the ionization energy is small), generating Li^+ ion. Similarly sodium ^{11}Na $(1s^2\ 2s^2\ 2p^6\ 3s)$ has a lone electron in the 3s orbital, and it is easily ionized just like ^3Li. This explains a similarity between the atoms (^3Li, ^{11}Na) located in the same column of the periodic table.

Let us consider a neutral atom containing Z electrons and a nucleus having the charge Ze. The cause of the binding is the coulomb attraction between the electrons and the nucleus plus the coulomb repulsion among the electrons. In the *self-consistent mean field approximation*, each electron moves in the medium of the total charge e concentrated at the nucleus similar to the electron in the hydrogen atom. Applying the Pauli exclusion principle to the electrons and considering the spin degeneracy, we can explain the periodic nature of the atomic binding (ionization) energy and the reason why there are eight columns in the periodic table. The hydrogenlike orbital description is therefore justified.

8.2 Fermi Liquid Model

We consider a monovalent metal, whose Hamiltonian H_A is

$$\begin{aligned} H_A &= \sum_{j=1}^{N} \frac{p_j^2}{2m} + \sum_{j>k}\sum \frac{k_0 e^2}{|\mathbf{r}_j - \mathbf{r}_k|} + \sum_{\alpha=1}^{N} \frac{P_\alpha^2}{2M} \\ &\quad + \sum_{\alpha>\gamma}\sum \frac{k_0 e^2}{|\mathbf{R}_\alpha - \mathbf{R}_\gamma|} - \sum_j \sum_\alpha \frac{k_0 e^2}{|\mathbf{r}_j - \mathbf{R}_\alpha|}. \end{aligned} \tag{8.1}$$

The motion of the set of N electrons is correlated because of the interelectronic coulomb interaction (the second sum). If we omit the ionic kinetic energy (the third sum) and the interelectronic and interionic coulomb interaction (the fourth sum) from Equation (8.1), we obtain

$$H_B = \sum_j \frac{p_j^2}{2m} + \sum_j V(\mathbf{r}_j) + \text{constant}, \tag{8.2}$$

$$V(\mathbf{r}) \equiv - \sum_\alpha \frac{k_0 e^2}{|\mathbf{r} - \mathbf{R}_\alpha|}, \tag{8.3}$$

which characterizes a system of free electrons moving in the *bare lattice*.

Since the metal as a whole is neutral, the coulomb interaction among the electrons, among the ions, and between electrons and ions, all have the same orders of magnitude, and hence they are equally important. We now pick one electron in the system. This electron is in interaction with the system of N ions and $N-1$ electrons, the environment (medium) having the net charge $+e$. The other $N-1$ electrons should, in accordance with Bloch's theorem, be distributed with the lattice periodicity and throughout the crystal in equilibrium. The charge per lattice ion is greatly reduced from e to $N^{-1}e$ because the net charge e of the medium is shared equally by N ions. As N is a large number, the selected electron moves in an *extremely weak effective lattice potential* V_e as characterized by the model Hamiltonian

$$h_C = \frac{p^2}{2m} + V_e(\mathbf{r}), \qquad V_e(\mathbf{r}+\mathbf{R}) = V_e(\mathbf{r}). \tag{8.4}$$

In other words, any chosen electron moves in an environment far different from what is represented by the bare lattice potential V. It moves almost freely in an extremely weak effective lattice potential V_e. This picture was obtained with the aid of Bloch's theorem, and hence it is a result of quantum theory. To further illustrate this, let us examine the same system from the classical point of view. In equilibrium the classical electron distribution is lattice-periodic; so there is one electron near each ion. However, the electron will not move in the greatly reduced field.

We now assume that the electrons move independently in the effective potential field V_e. The total Hamiltonian for the idealized system may then be represented by

$$H_C = \sum_j h_C(\mathbf{r}_j, \mathbf{p}_j) \equiv \sum_j \left[\frac{p_j^2}{2m} + V_e(\mathbf{r}_j)\right]. \tag{8.5}$$

This Hamiltonian H_C is a far better approximation to the original Hamiltonian H_A in Equation (8.1) than the Hamiltonian H_B in Equation (8.2). In H_C both interelectronic and interionic coulomb repulsions are not neglected but are taken into consideration self-consistently. This model is a *one-electron-picture approximation*, but it is hard to improve on by any simple method. The model, in fact, forms the basis for the band theory of electrons.

We now apply Bloch's theorem to the Hamiltonian H_C composed of the kinetic energy and the interaction energy V_e. We then obtain the Bloch energy bands $\epsilon_j(\hbar\mathbf{k})$ and the Bloch states characterized by band index j and k-vector $\mathbf{k}$. The Fermi–Dirac statistics obeyed by the electrons can be applied to the Bloch electrons with no regard to the interaction. There is a certain Fermi energy ϵ_F for the ground state of the system. Thus, there is a *sharp*

Fermi surface represented by

$$\epsilon_j(\hbar\mathbf{k}) = \epsilon_F, \tag{8.6}$$

which separates the electron-filled k-space (low-energy side) from the empty k-space (high-energy side).

The Fermi surface for a real metal in general is complicated in contrast to the free-electron Fermi sphere represented by

$$\frac{p^2}{2m} \equiv \frac{p_x^2 + p_y^2 + p_z^2}{2m} = \epsilon_F. \tag{8.7}$$

The independent electron model with a sharp Fermi surface at 0 K is called the *Fermi liquid model*. This model is experimentally supported, as we see later in Section 9.2, by the low-temperature linear heat capacity observed in metals.

The Fermi liquid model was obtained in the static lattice approximation where the motion of the ions is neglected. If the effect of moving ions (phonons) is taken into account, a new Hamiltonian is required. The electron-phonon interaction turns out to be very important in the theory of superconductivity.

Chapter 9

The Fermi Surface

At 0 K the normal metal has a sharp Fermi surface, which is experimentally supported by the fact that the heat capacity is linear in the temperature T at the lowest temperatures. The Fermi surfaces in a few metals are discussed in this chapter.

9.1 Monovalent Metals (Na, Cu)

Why does a particlar metal exist with a particular crystalline state? This is a good question but very hard to answer. The answer must involve the composition and nature of the atoms constituting the metal and the interaction between the component particles. To illustrate these complications, let us take sodium (Na), which is known to form a BCC lattice. This monovalent metal may be thought of as an ideal composite system of electrons and ions interacting with the coulomb forces. The system Hamiltonian may be approximated by H_A in Equation (8.1), which consists of the kinetic energies of electrons and ions and the coulomb interaction energies among and between the electrons and ions. This is an approximation because the interaction between the electron and the ion deviates significantly from the ideal coulomb law at short distances because each ion has core electrons. The study of the ground-state energy of the ideal model favors a FCC lattice structure, which is not observed for this metal. If multivalent metals like lead (Pb) are considered, the situation becomes even more complicated since the core electrons forming part of the ions have anisotropic charge distribution. Therefore, it is customary in solid-state physics to assume the experimentally known lattice structures first, then proceed to study the Fermi surface.

Once a lattice is selected, the Brillouin zone is fixed. For an ORC lattice the Brillouin zone is a rectangular box defined by Equation (7.26). Let us consider a large periodic box of volume

$$V = (N_1 a)(N_2 b)(N_3 c), \qquad N_1,\, N_2,\, N_3 \gg 1. \tag{9.1}$$

Let us find the number N of the quantum states within each Brillouin zone. If we ignore the spin degeneracy, the number N is equal to the total phase-space volume divided by unit phase-cell volume:

$$\left(\frac{2\pi\hbar}{a}\right)\left(\frac{2\pi\hbar}{b}\right)\left(\frac{2\pi\hbar}{c}\right) \div \left(\frac{2\pi\hbar}{N_1 a}\right)\left(\frac{2\pi\hbar}{N_2 b}\right)\left(\frac{2\pi\hbar}{N_3 c}\right) = N_1 N_2 N_3. \tag{9.2}$$

The spin degeneracy doubles this number. We note that the product $N_1 N_2 N_3$ simply equals the number of the ions in the normalization volume. It is also equal to the number of the conduction electrons in a monovalent metal. In other words, *the first Brillouin zone for a monovalent metal can contain twice the number of conduction electrons*. This means that at 0 K, half of the Brillouin zone may be filled by electrons. For example, if an energy-momentum relation Equation (7.42) is assumed, the Fermi surface is an ellipsoid. Something similar to this occurs with alkali metals including Li, Na, and K. These metals form BCC lattices. All experiments indicate that their Fermi surfaces are spherical and are entirely within the first Brillouin zone. The Fermi surface of sodium is shown in Figure 9.1.

The *nearly free electron model* (NFEM) developed by Harrison [1] can predict a Fermi surface in the first approximation for any metal. (Interested readers are referred to the book by Harrison [2].) This model is obtained by applying the Heisenberg uncertainty principle and Pauli exclusion principle to a solid, and therefore it has a general applicability unhindered by the complications due to particle-particle interaction. Briefly in the NFEM, the first Brillouin zone is drawn for a selected metal. Electrons are filled starting from the center of the zone, with the assumption of a free-electron energy-momentum relation [Equation (7.41)]. If we apply the NFEM to alkali metals, we simply obtain the Fermi sphere shown in Figure 9.1.

Noble metals, including copper (Cu), silver (Ag), and gold (Au), are monovalent FCC metals. The Brillouin zone and Fermi surface of copper are shown in Figure 9.2. The Fermi surface is far from spherical. (How such a Fermi surface is obtained after the analyses of the de Haas–van Alphen effect

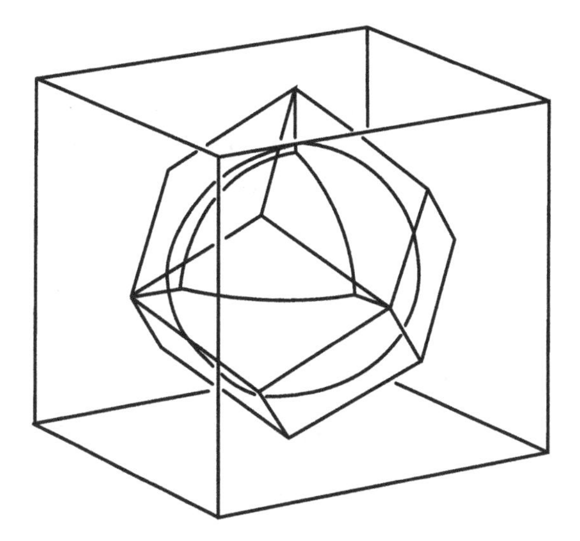

Figure 9.1: The Fermi surface of sodium (BCC) is nearly spherical within the first Brillouin zone.

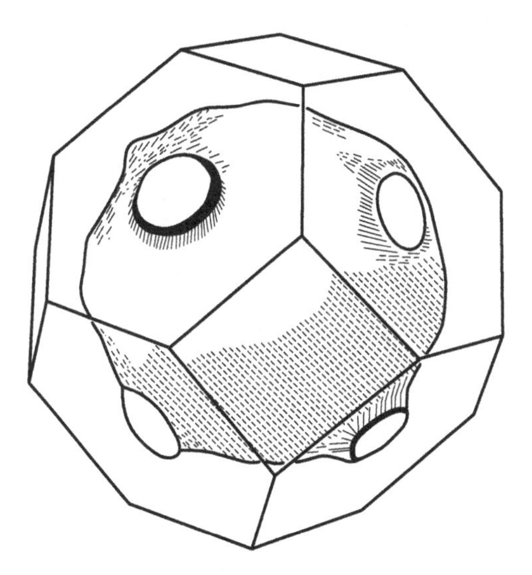

Figure 9.2: The Fermi surface of copper (FCC). The electron sphere bulges out in the ⟨111⟩ direction to make contact with the hexagonal zone faces at right angles.

will be discussed in Section 9.3.)

Notice that the *Fermi surface approaches the Brillouin boundary at right angles*. This arises from the *mirror symmetry* possessed by the FCC lattice, which will be shown here. From Equation (7.24), we obtain

$$\psi_E(\mathbf{r} + \mathbf{R}) = e^{i(\mathbf{k}+\mathbf{G})\cdot\mathbf{R}}\psi_E(\mathbf{r}), \tag{9.3}$$

where $\mathbf{G}$ is any reciprocal lattice vector satisfying

$$\exp(i\mathbf{G} \cdot \mathbf{R}) = 1. \tag{9.4}$$

Equation (9.3) means that the energy $E = \epsilon_j(\hbar\mathbf{k})$ is periodic in the k-space:

$$\epsilon_j(\hbar\mathbf{k} + \hbar\mathbf{G}) = \epsilon_j(\hbar\mathbf{k}). \tag{9.5}$$

If the lattice has a mirror symmetry, we can choose the symmetry plane as the yz-plane and express the symmetry by

$$V(x, y, z) = V(-x, y, z). \tag{9.6}$$

The Schrödinger equation (7.23) is invariant under the mirror reflection $(x \longrightarrow -x)$. This means that $\psi_E(x)$ and $\psi_E(-x)$ satisfy the same equation, where we omitted the y- and z-dependence. Using the Bloch theorem, Equation (7.24), we obtain

$$\psi_E(-x - a) = e^{-ika}\psi_E(-x). \tag{9.7}$$

Taking a mirror reflection of this equation, we obtain

$$\psi_E(x + a) = e^{-ika}\psi_E(x). \tag{9.8}$$

Combination of Equations (7.24) and (9.8) means that the energy $\epsilon_j(k)$ is an even function of k:

$$\epsilon_j(-k) = \epsilon_j(k). \tag{9.9}$$

The opposing faces of the Brillouin boundary are often separated by the fundamental reciprocal-lattice constants, and the mirror symmetry planes are located in the middle of these faces. This is true for many familiar lattices, including FCC, BCC, SC, diamond, ORC, hexagonal closed-pack (HCP). Differentiating Equation (9.9) with respect to k, we obtain

$$\epsilon_j'(k) \equiv \frac{d}{dk}\epsilon_j(k) = -\epsilon_j'(-k). \tag{9.10}$$

By setting $k = \pi/a$, we have $\epsilon_j'(\pi/a) = -\epsilon_j'(-\pi/a) = -\epsilon_j'(\pi/a)$, where the last equality follows from Equation (9.9). We then obtain

$$\epsilon_j'\left(\frac{\pi}{a}\right) = 0. \qquad \text{QED} \tag{9.11}$$

For a divalent metal like calcium (Ca) (FCC), the first Brillouin zone can in principle contain all of the conduction electrons. However, the Fermi surface must approach the zone boundary at right angles, which distorts the ideal configuration considerably. Thus, the real Fermi surface for Ca has a set of unfilled corners in the first zone, and the overflow electrons are in the second zone. As a result Ca is a metal and not an insulator.

9.2 Multivalent Metals

Divalent beryllium (Be) forms an HCP crystal. The Fermi surfaces in the second zone, (a) constructed in the NFEM and (b) observed [3], are shown in Figure 9.3. The reason why the "monster" in Figure 9.3 (a), which has sharp corners, must be smoothed out will be explained later in Section 10.2.

Transition metal tungsten (W), which contains d-electrons [(xenon) $4f^{14}5d^46s^2$], forms a BCC crystal. The conjectured Fermi surfaces are shown in Figure 9.4 [3].

Let us now consider trivalent aluminum (Al), which forms a FCC lattice. The first Brillouin zone is completely filled with electrons. The second zone is half-filled with electrons, starting with the zone boundary as shown in Figure 9.5. We examine the Fermi surface of lead (Pb), which also forms a FCC lattice. Since this metal is quadrivalent and, therefore, has a great number of conduction electrons, the Fermi surface is quite complicated. The conjectured Fermi surface in the third zone is shown in Figure 9.6. (For more detailed description of the Fermi surfaces for metals see standard texts on solid-states physics [2, 4].) Al, Be, W, and Pb are superconductors, whereas monovalent Na and Cu are not. This difference will be discussed later.

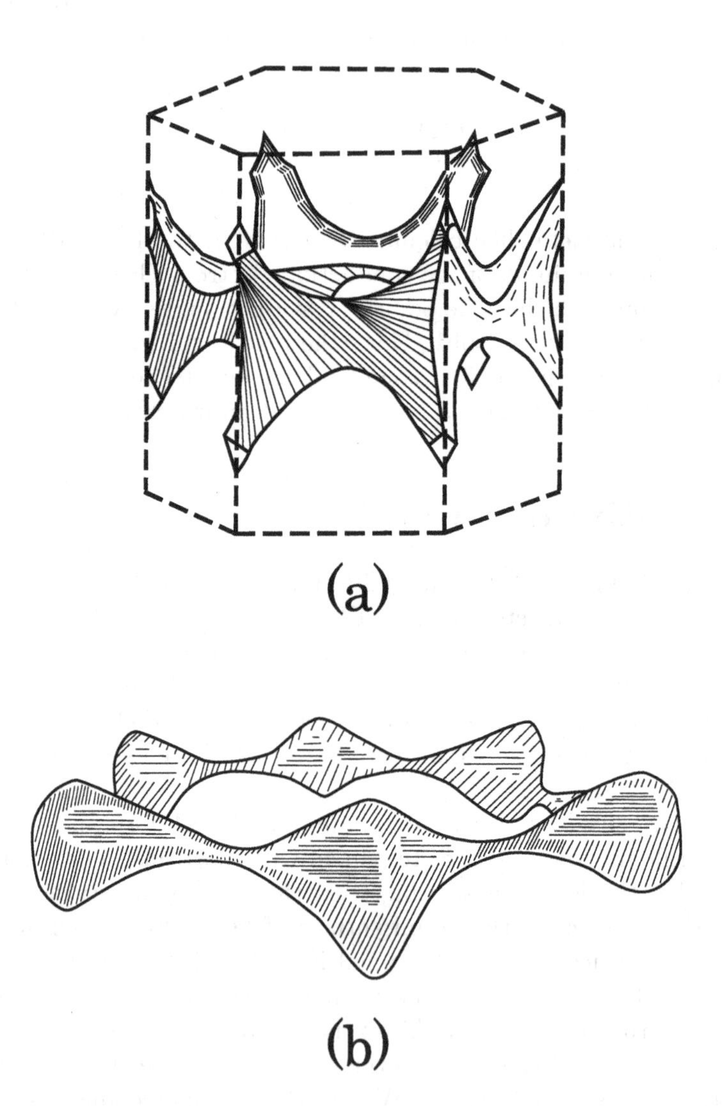

Figure 9.3: The Fermi surfaces in the second zone for beryllium (Be): (a) NFEM monster, (b) measured coronet. The coronet encloses unoccupied states.

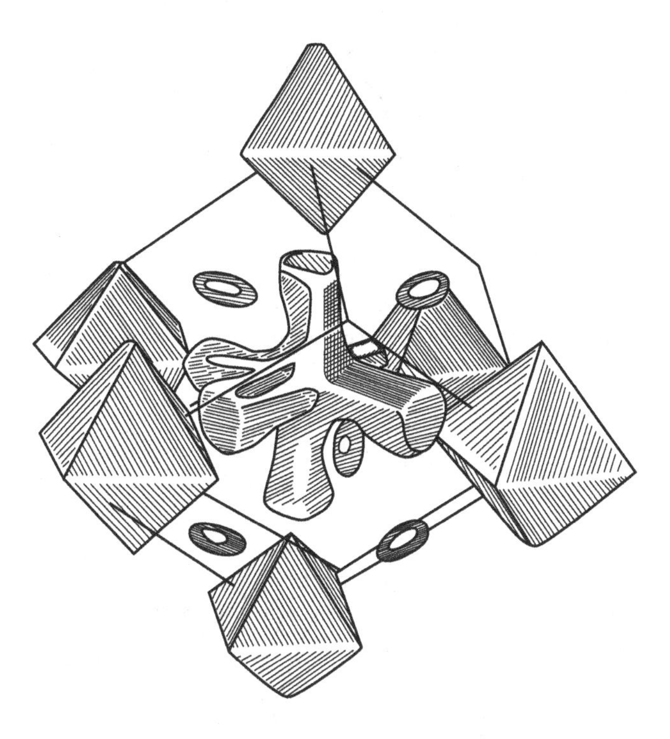

Figure 9.4: The Fermi surfaces for tungsten (BCC). The central figure contains electrons, and all other figures contain vacant states (after Schönberg and Gold [3]).

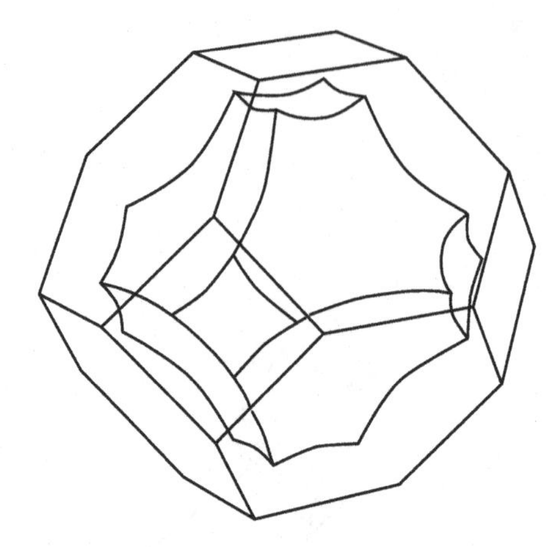

Figure 9.5: The Fermi surface constructed by Harrison's NFEM model in the second zone for aluminum (Al). The convex surface encloses vacant states.

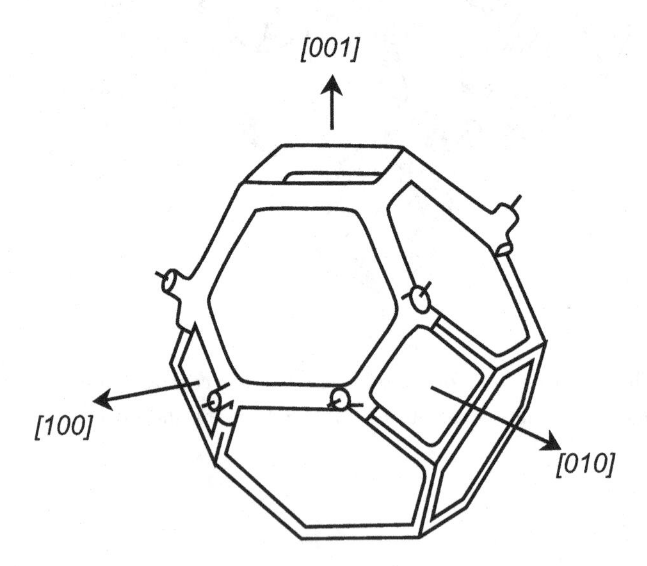

Figure 9.6: The conjectured Fermi surface in the third zone for lead (FCC). The cylinder encloses electrons.

9.3 Electronic Heat Capacity and Density of States

The band structures of conduction electrons are different from metal to metal. However, the electronic heat capacities for conduction electrons at very low temperatures are all similar, which is shown in this section.

In Sections 3.3 and 3.4, we saw that the heat capacity of free electrons has a T-linear dependence at low temperatures (below 1 K). We first show that *any normal metal having a sharp Fermi surface must have a T-linear heat capacity*. Let us follow the approximate calculations described in Section 3.3. The number of excited electrons N_X is estimated by [see Figure 3.5 and Equation (3.39)]

$$N_X = \mathcal{N}(\epsilon_F) k_B T, \tag{9.12}$$

where $\mathcal{N}(\epsilon_F)$ is the density of states at the Fermi energy ϵ_F. Each thermally excited electron will move up with extra energy of the order $k_B T$. The approximate change in the total energy ΔE is given by multiplying these two factors:

$$\Delta E = N_x k_B T = \mathcal{N}(\epsilon_F)(k_B T)^2. \tag{9.13}$$

Differentiating this equation with respect to T, we obtain an expression for the heat capacity:

$$C_V \cong \frac{\partial}{\partial T} \Delta E = 2k_B^2 \mathcal{N}(\epsilon_F) T, \tag{9.14}$$

which indicates the T-linear dependence. This temperature dependence comes from the Fermi distribution function. We can improve our calculation by following Sommerfeld's method outlined in Section 3.4. Comparing Equation (3.43) with Equation (3.60), the correction factor is $(\pi^2/2)/3 = \pi^2/6$. Using this correction factor, we obtain from Equation (9.14)

$$\boxed{C_V = \left(\frac{1}{3}\right) \pi^2 k_B^2 \mathcal{N}(\epsilon_F) T,} \tag{9.15}$$

which allows quantitative comparisons with experiments, since the coefficient $[\pi^2 k_B^2 \mathcal{N}(\epsilon_F)/3]$ is calculated exactly.

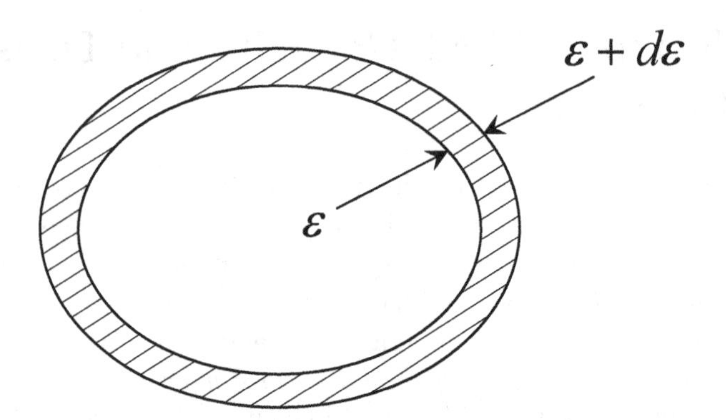

Figure 9.7: Two constant-energy curves at ϵ and $\epsilon + d\epsilon$ in the two-dimensional k-space.

The density of states $\mathcal{N}(\epsilon_F)$ for any 3D normal metal can be expressed by

$$\boxed{\mathcal{N}(\epsilon_F) = \sum_j \frac{2}{(2\pi\hbar)^3} \int dS \, \frac{1}{|\nabla \epsilon_j(\hbar \mathbf{k})|},} \tag{9.16}$$

where 2 is the spin degeneracy factor and the surface integration is carried out over the Fermi surface represented by

$$\epsilon_j(\hbar \mathbf{k}) = \epsilon_F. \tag{9.17}$$

Equation (9.16) may be proved as follows. First, consider the 2D case. Assume that the electron moves on the xy-plane only. The constant energy surface is then a closed curve in the $k_x k_y$-plane. We choose two energy surfaces at ϵ and $\epsilon + d\epsilon$ as shown in Figure 9.7. Twice the area between the two curves divided by the unit quantum cell area $(2\pi\hbar)^2$ equals the number of the quantum states in the range (ϵ, $\epsilon + d\epsilon$). The density of states per unit area $\mathcal{N}(\epsilon)$ is then determined from

$$\mathcal{N}(\epsilon) d\epsilon = \frac{2}{(2\pi\hbar)^2} (\text{shaded area}) = \frac{2\hbar}{(2\pi\hbar)^2} \int_C dk \, \frac{d\epsilon}{|\nabla \epsilon(\hbar \mathbf{k})|}, \tag{9.18}$$

where the integration is carried out along the constant-energy curve C represented by $\epsilon(\hbar \mathbf{k}) = \epsilon$. The equality between the last two members of Equations

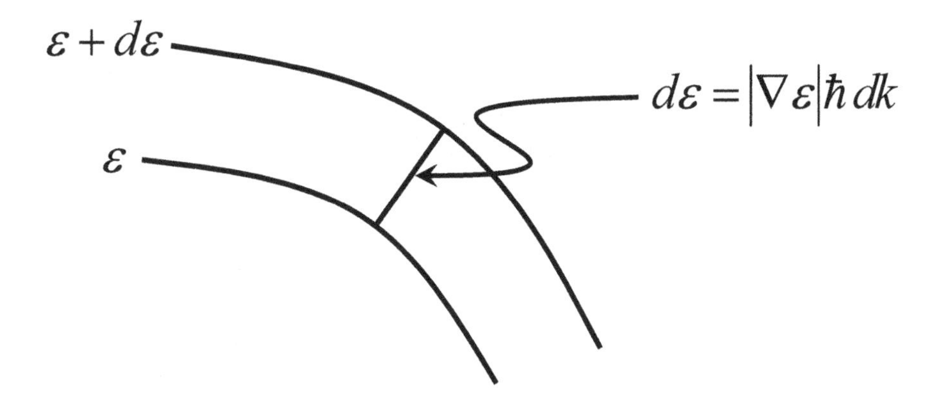

Figure 9.8: A diagram indicating the relation $d\epsilon = |\nabla\epsilon(\hbar\mathbf{k})|\hbar dk$.

(9.18) can be seen from the k-space (plane) diagram shown in Figure 9.8:

$$d\epsilon = |\nabla\epsilon(\hbar\mathbf{k})|\hbar dk. \tag{9.19}$$

This graphic method can simply be extended to three dimensions. If the Fermi surface extends over two zones, the total density of states is the sum over the zones (j). QED.

For example, consider a free-electron system having the Fermi sphere

$$\epsilon = (p_x^2 + p_y^2 + p_z^2)/(2m) \equiv \epsilon_F. \tag{9.20}$$

The gradient $\nabla\epsilon(\mathbf{p})$ at any point of the surface has a constant magnitude p_F/m, and the surface integral is equal to $4\pi p_F^2$. Equation (9.16) then yields

$$\mathcal{N}(\epsilon_F) = V\left[2(2\pi\hbar)^{-3}\right]\left[\frac{4\pi p_F^2}{(p_F/m)}\right] = V\left[\frac{2^{1/2}m^{3/2}}{\pi^2\hbar^3}\right]\epsilon_F^{1/2}, \tag{9.21}$$

in agreement with Equation (3.36). For a second example, consider the ellipsoidal surface represented by Equation (7.42). After elementary calculations, we obtain (Problem 9.3.1)

$$\mathcal{N}(\epsilon) = V\left(\frac{2^{1/2}}{\pi^2\hbar^3}\right)(m_1m_2m_3)^{1/2}\epsilon^{1/2}, \tag{9.22}$$

which shows that the density of states still grows like $\epsilon^{1/2}$, but the coefficient depends on the three effective masses (m_1, m_2, m_3).

Problem 9.3.1.

a. Compute the momentum-space volume between the surfaces represented by $\epsilon = p_x^2/2m_1 + p_y^2/2m_2 + p_z^2/2m_3$ and $\epsilon + d\epsilon = p_x'^2/2m_1 + p_y'^2/2m_2 + p_z'^2/2m_3$. By counting the number of quantum states in this volume in the bulk limit, obtain Equation (9.22).

b. Derive Equation (9.22), starting from the general formula (9.16). Hint: Convert the integral over the ellipsoidal surface to one over a spherical surface.

Chapter 10

Bloch Electron Dynamics

Newtonian equations of motion for the Bloch electron are derived and discussed in this chapter. "Electrons" ("holes"), which appear in the Hall coefficient mesurements, are generated near the Fermi surface on the negative (positive) curvature side of the surface.

10.1 Introduction

First, we briefly state the reasons why we need new *Newtonian equations of motion* for the Bloch electron.

Wilson, in his well-known book *Theory of Metals* [1], wrote down the equations of motion for a Bloch electron (wave packet) as

$$\sum_j m_{ij} \frac{dv_j}{dt} = qE_i, \tag{10.1}$$

where $\mathbf{E}$ is the electric field;

$$\mathbf{v} \equiv \dot{\mathbf{r}} = \frac{\partial \epsilon_j(\mathbf{p})}{\partial \mathbf{p}}, \qquad \mathbf{p} \equiv \hbar \mathbf{k}, \tag{10.2}$$

is the velocity of the electron with charge $q = -e$. Here, $\epsilon = \epsilon_j(\mathbf{p}) \equiv \epsilon_j(\hbar\mathbf{k})$ is the energy eigenvalue characterized by the wave vector $\mathbf{k}$ and the zone number j associated with the Schrödinger equation

$$\left[-\frac{\hbar^2}{2m}\nabla^2 + V_e(\mathbf{r})\right] \psi_{j,k}(\mathbf{r}) = \epsilon\, \psi_{j,k}(\mathbf{r}) \tag{10.3}$$

with V_e representing an effective lattice potential (see Section 8.2), and $\{m_{ij}\}$ are elements of a symmetric *mass tensor* $\mathcal{M}$, whose inverse are defined by

$$\left(\mathcal{M}^{-1}\right)_{ij} \equiv \frac{\partial^2 \epsilon}{\partial p_i \partial p_j}. \tag{10.4}$$

According to special relativity theory, there is no separate force on a charge; any charged particle is simply acted on by the electric force acting in its own frame of reference in which $\mathbf{v} = 0$, and the action of a magnetic force can be determined by an appropriate Lorentz transformation [2]. Consequently, the electric and magnetic fields $(\mathbf{E}, \mathbf{B})$ act in a package as the *Lorentz force*. We may, therefore, modify Equation (10.1) to

$$\sum_j m_{ij} \frac{dv_j}{dt} = q(\mathbf{E} + \mathbf{v} \times \mathbf{B})_i, \qquad i = 1, 2, 3. \tag{10.5}$$

Strictly speaking, the lhs should have a relativistic correction, which is negligibly small since the electron speed v is much lower than the speed of light. In many other textbooks and monographs [3–7], there is a second equation,

$$\hbar \frac{d\mathbf{k}}{dt} = q(\mathbf{E} + \mathbf{v} \times \mathbf{B}), \tag{10.6}$$

which is often called a *semiclassical equation of motion*.

Equation (10.5) are Newtonian equations of motion; they describe how the velocity $\mathbf{v}$ of the Bloch electron (wave packet) changes by the Lorentz force. They are clearly gauge invariant. No scalar and vector potentials appear here. The meaning of Equation (10.6) is not transparent; as it stand, it appears to indicate that the quantum k-vector changes by the action of the Lorentz force, which is not in accord with the standard quantum laws of motion in either the Heisenberg or the Schrödinger picture. A serious difficulty of Equation (10.6) arises when we consider inhomogeneous electromagnetic fields $[\mathbf{E}(\mathbf{r}), \mathbf{B}(\mathbf{r})]$. Since the vector $\mathbf{k}$ is Fourier conjugate to the position $\mathbf{r}$, it cannot depend on this position $\mathbf{r}$. Hence, Equation (10.6) may be valid for homogeneous fields only, which is a severe limitation. The solution of Equation (10.6) indicates a further limitation. Consider the case of $\mathbf{B} = 0$. Assuming $\mathbf{E} = (E, 0, 0)$, we have a solution: $\hbar k_x = qEt + p_0$, where p_0 is a constant. This solution does not depend on the energy-momentum relation. It is material- and orientation-independent. Therefore, we cannot discuss the anisotropic electron motion in layered conductors like graphite. We shall show that Equation (10.5) is the correct one to use in Bloch electron dynamics.

10.2 Newtonian Equations of Motion

In Section 9.2 we saw that the electronic heat capacity is generated by the thermal excitation of the electrons near the Fermi surface. These same electrons also participate in the charge transport. Here we discuss how these electrons respond to the applied electromagnetic fields.

Let us recall that in the Fermi liquid model each electron in a crystal moves independently in an extremely weak lattice-periodic effective potential $V_e(\mathbf{r})$:

$$V_e(\mathbf{r} + \mathbf{R}) = V_e(\mathbf{r}). \tag{10.7}$$

We write down the Schrödinger equation corresponding to the single-electron Hamiltonian h_c in Equation (8.4):

$$\left[-\frac{\hbar^2}{2m}\nabla^2 + V_e(\mathbf{r})\right]\psi(\mathbf{r}) = E\psi(\mathbf{r}). \tag{10.8}$$

According to Bloch's theorem, the wave function ψ satisfies

$$\psi_{j,\mathbf{k}}(\mathbf{r} + \mathbf{R}) = e^{i\mathbf{k}\cdot\mathbf{R}}\,\psi_{j,\mathbf{k}}(\mathbf{r}). \tag{10.9}$$

The Bravais vector $\mathbf{R}$ can take on only discrete values, and its minimum length is equal to the lattice constant a. This generates a limitation on the domain in $\mathbf{k}$. For example, the values for each k_a ($a = x,\ y,\ z$) for a SC lattice are limited to $(-\pi/a,\ \pi/a)$. This means that the Bloch electron's wavelength $\lambda \equiv 2\pi/k_a$ has a lower bound:

$$\lambda > 2a. \tag{10.10}$$

The Bloch electron state is characterized by k-vector $\mathbf{k}$, band index j, and energy

$$\epsilon = \epsilon_j(\hbar\mathbf{k}) \equiv \epsilon_j(\mathbf{p}). \tag{10.11}$$

The *energy-momentum relation*, also called the *dispersion relation*, represented by Equation (10.11) can be probed by transport measurements. A metal is perturbed from the equilibrium condition by an applied electric field; the deviations of the electron distribution from the equilibrium move in the crystal to reach equilibrium and maintain a stationary state. The deviations, that is, the *localized Bloch wave packets*, should extend over several lattice units. This is so because no wave packets constructed from waves of the k-vectors (k_x, k_y, k_z) whose magnitudes have the upper bounds (π/a) can be localized within distances less than a.

Dirac demonstrated in his famous book, *Principles of Quantum Mechanics* [8], that for any p dependence of the kinetic energy $[\epsilon \equiv \epsilon_j(\mathbf{p})]$ the *center of a wave packet*, identified as the position of the quantum particle, moves in accordance with Hamilton's equations of motion. Thus *the Bloch electron representing the wave packet should move classical mechanically under the action of the force averaged over the lattice constants.* The *lattice force* $-\partial V_e/\partial x$ averaged over a unit cell vanishes:

$$\left\langle -\frac{\partial}{\partial x} V_e \right\rangle_{\text{unit cell}} \equiv -a^{-3} \int\int dydz \int_0^a dx \frac{\partial}{\partial x} V_e(x, y, z) = 0. \tag{10.12}$$

The only important forces acting on the Bloch electron are electromagnetic forces.

We now formulate a Bloch electron dynamics as follows [9]. First, from the quantum principle of *wave-particle duality*, we postulate that

$$(\hbar k_x, \hbar k_y, \hbar k_z) = (p_x, p_y, p_z) \equiv (p_1, p_2, p_3) \equiv \mathbf{p}. \tag{10.13}$$

Second, we introduce a model Hamiltonian,

$$H_0(p_1, p_2, p_3) \equiv \epsilon_j(p_1, p_2, p_3). \tag{10.14}$$

Third, we generalize our Hamiltonian H to include the electromagnetic interaction energy,

$$H = H_0(\mathbf{p} - q\mathbf{A}) + q\phi, \tag{10.15}$$

where $(\mathbf{A}, \phi)$ are vector and scalar potentials generating electromagnetic fields $(\mathbf{E}, \mathbf{B})$:

$$\mathbf{E} = -\nabla\phi(\mathbf{r}, t) - \frac{\partial \mathbf{A}(\mathbf{r}, t)}{\partial t}, \quad \mathbf{B} = \nabla \times \mathbf{A}(\mathbf{r}, t), \quad \mathbf{r} \equiv (x_1, x_2, x_3). \tag{10.16}$$

By using the standard procedures, we then obtain Hamilton's equations of motion:

$$\dot{\mathbf{r}} \equiv \mathbf{v} = \frac{\partial H}{\partial \mathbf{p}} = \frac{\partial}{\partial \mathbf{p}} H_0(\mathbf{p} - q\mathbf{A}), \tag{10.17}$$

$$\dot{\mathbf{p}} = -\frac{\partial H}{\partial \mathbf{r}} = -\frac{\partial}{\partial \mathbf{r}} H_0 - q\frac{\partial}{\partial \mathbf{r}}\phi. \tag{10.18}$$

Equation (10.17) defines the velocity $\mathbf{v} \equiv (v_1, v_2, v_3)$. Notice that in the zero-field limit these equations are in agreement with the general definition of a *group velocity*:

$$v_{g,i} \equiv \frac{\partial \omega(\mathbf{k})}{\partial k_i}, \quad \omega(\mathbf{k}) \equiv \frac{\epsilon(\mathbf{p})}{\hbar} \quad \text{(wave picture)},$$

$$v_i \equiv \frac{\partial \epsilon(\mathbf{p})}{\partial p_i} \quad \text{(particle picture)}. \tag{10.19}$$

In the presence of a magnetic field, Equation (10.17) gives the velocity $\mathbf{v}$ as a function of $\mathbf{p} - q\mathbf{A}$. Inverting this functional relation, we have

$$\mathbf{p} - q\mathbf{A} = \mathbf{f}(\mathbf{v}). \tag{10.20}$$

Using Equation (10.20) and Equation (10.18), we obtain (Problem 10.1.2)

$$\frac{d\mathbf{f}}{dt} = q(\mathbf{E} + \mathbf{v} \times \mathbf{B}). \tag{10.21}$$

Since the vector $\mathbf{f}$ is a function of $\mathbf{v}$, Equation (10.21) describes how the velocity $\mathbf{v}$ changes by the action of the Lorentz force (the right-hand term).

To see the nature of Equation (10.21), let us take a quadratic dispersion relation represented by

$$\epsilon = \frac{p_1^2}{2m_1^*} + \frac{p_2^2}{2m_2^*} + \frac{p_3^2}{2m_3^*}, \tag{10.22}$$

where $\{m_j^*\}$ are *effective masses*. The effective masses $\{m_j^*\}$ may be positive or negative. Depending on their values the energy surface represented by Equation (10.22) is ellipsoidal or hyperboloidal (see Figure 10.1). If the Cartesian axes are taken along the major axes of the ellipsoid, Equation (10.21) can be written as (Problem 10.1.1)

$$m_j^* \frac{dv_j}{dt} = q(\mathbf{E} + \mathbf{v} \times \mathbf{B})_j. \tag{10.23}$$

These are *Newtonian equations of motion*: mass $\times$ acceleration $=$ force. Only a set of three effective masses (m_j^*) is introduced. Thus, the Bloch electron moves in an anisotropic environment if the effective masses are different.

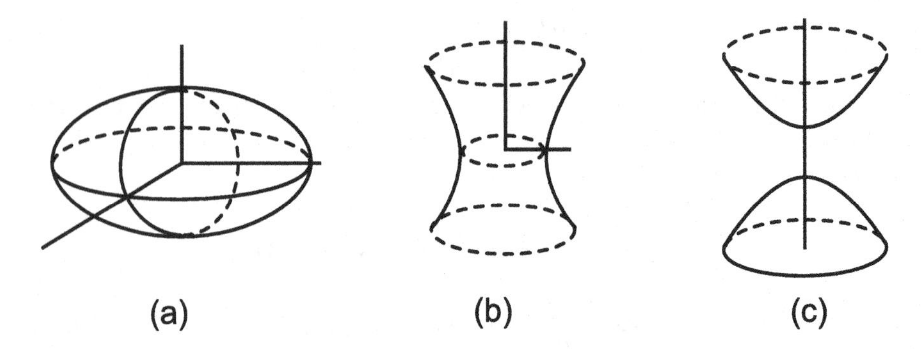

Figure 10.1: (a)Ellipsoid, (b)hyperboloid of one sheet (neck) and (c) hyperboloid of two sheets (inverted double caps).

Let us now go back to the general case. The function **f** may be determined from the dispersion relation (10.11) as follows. Take a point A at the *constant-energy surface* represented by Equation (10.11) in the k-space. We choose the *positive* normal vector to point in the direction of increasing energy. A *normal curvature* κ is defined to be the inverse of the radius of the contact circle at A, R_A, in the plane containing the normal vector times the sign of curvature δ_A:

$$\kappa \equiv \delta_A R_A^{-1}. \tag{10.24}$$

The *sign of the curvature*, δ_A, is $+1$ or -1 according to whether the center of the contact circle is on the positive side (which contains the positive normal) or not. In space-surface theory (e.g., see Ref. 10), it is known that the two planes which contain the greatest and smallest normal curvatures are mutually orthogonal. These planes are by construction orthogonal to the contact plane at A. Thus, the intersections of these two planes and the contact plane form a Cartesian set of orthogonal axes with the origin at A, called the *principal axes of curvatures*. By using this property, we define *principal masses* m_i by

$$\boxed{\frac{1}{m_i} \equiv \frac{\partial^2 \epsilon}{\partial p_i^2},} \tag{10.25}$$

where dp_i is the differential along the principal axis i. If we choose a Cartesian coordinate system along the principal axes, Equation (10.21) can, then, by written as

$$\boxed{m_i \frac{dv_i}{dt} = q(\mathbf{E} + \mathbf{v} \times \mathbf{B})_i.} \tag{10.26}$$

These equations are similar to the equation of motion Equation (10.23). The principal masses $\{m_i\}$, however, are defined at each point on the constant-energy surface, and hence they depend on $\mathbf{p}$ and $\epsilon_j(\mathbf{p})$. Let us take a simple example: the ellipsoidal constant-energy surface represented by Equation (10.22) with all positive m_i^*. At extremal points, e.g., $(p_{1,\max}, 0, 0) = ([2\epsilon m_1^*]^{1/2}, 0, 0)$, the principal axes of curvatures match the major axes of the ellipsoid. Then the principal masses $\{m_i\}$ can simply be expressed in terms of the constant effective masses $\{m_j^*\}$ (Problem 10.1.3).

Let us now derive Newtonian equations valid in *any* Cartesian frame of reference, starting with Equation (10.21). As f_j are functions of (v_1, v_2, v_3), we obtain

$$\frac{df_i}{dt} = \sum_j \frac{\partial f_i}{\partial v_j}\frac{dv_j}{dt} = \sum_j \frac{1}{\partial v_j/\partial f_i}\frac{dv_j}{dt}. \tag{10.27}$$

The velocities v_i from Equation (10.17) can be expressed in terms of the first p-derivatives. Thus, in the zero-field limit we obtain

$$\boxed{\frac{\partial v_j}{\partial f_i} \to \frac{\partial^2 \epsilon}{\partial p_i \partial p_j} \equiv \frac{1}{m_{ij}},} \tag{10.28}$$

which defines the *mass tensor elements* $\{m_{ij}\}$. By using Equations (10.27) and (10.28), we can re-express Equation (10.21) as

$$\boxed{\sum_j m_{ij} \frac{dv_j}{dt} = q(\mathbf{E} + \mathbf{v} \times \mathbf{B})_i,} \tag{10.29}$$

which are in agreement with Wilson's equations (10.5).

The mass tensor $\{m_{ij}\}$ in Equation (10.28) is *real* and *symmetric*:

$$m_{ij} = m_{ji}. \tag{10.30}$$

Any real symmetric tensor can always be diagonalized by a *principal-axes transformation* [11]. The principal masses $\{m_i\}$ are given by Equation (10.25), and the principal axes are given by the principal axes of curvature.

In Equation (10.25) the third principal mass m_3 is defined in terms of the second derivative $\partial^2\epsilon/\partial p_3^2$ in the energy-increasing (p_3-) direction. The first and second principal masses (m_1, m_2) can be connected with the two principal radii of curvature (P_1, P_2), which by definition equal the inverses of the two principal curvatures (κ_1, κ_2) (Problem 10.1.4):

$$\frac{1}{m_j} = -\kappa_j v \equiv -\frac{v}{P_j}, \qquad v \equiv |\mathbf{v}|, \qquad \frac{1}{P_j} \equiv \kappa_j. \tag{10.31}$$

Equation (10.31) presents very useful relations. The signs (definitely) and magnitudes (qualitatively) of the first two principal masses (m_1, m_2) can be obtained by a visual inspection of the constant-energy surface. An example of a constant-energy surface is the Fermi surface. The sign of the third principal mass m_3 can also be obtained by inspection: the mass m_3 is positive or negative according to whether the center of the contact circle is on the negative or the positive side. For example, the system of free electrons has a spherical constant-energy surface represented by $\epsilon = p^2/(2m)$ with the normal vector pointing outward. By inspection the principal radii of curvatures at every point of the surface are negative, and therefore, according to Equation (10.31), the principal masses (m_1, m_2) are positive and equal to m. The third principal mass m_3 is also positive and equal to m.

Equation (10.26) was derived from the dispersion relation [Equation (10.11)] without referring to the Fermi energy. This equation is valid for all wave vectors $\mathbf{k}$ and all band indices j. Further discussion of Equation (10.26) will be given in Section 10.3.

Problem 10.2.1. Assume a quadratic dispersion relation [Equation (10.22)] and derive Equation (10.23).

Problem 10.2.2. Assume a general dispersion relation [Equation (10.14)] and derive Equation (10.26).

Problem 10.2.3. Consider the ellipsoidal constant-energy surface represented by Equation (10.22) with all $m_i^* > 0$. At the six extremal points, the principal axes of curvatures match the major axes of the ellipsoid. Demonstrate that the principal masses $\{m_i\}$ at one of these points can be expressed simply in terms of the effective masses $\{m_j^*\}$.

Problem 10.2.4. Verify Equation (10.31). Use a Taylor expansion.

10.3 Discussion

The velocity $\mathbf{v}$ in Equation (10.26) is defined in terms of the derivatives of the energy $\epsilon_j(\mathbf{p}) \equiv \epsilon_j(\hbar\mathbf{k})$, which are the energy eigenvalues of the zero-field Schrödinger equation (10.8). If a constant magnetic field is applied, the quantum states are characterized by the Landau quantum numbers $(n,\ k_y,\ k_z)$ (see Section 4.3), which are distinct from the Bloch quantum numbers $(j, \mathbf{k})$. The principal masses and effective masses are defined in the low-field limit. The validity of Equation (10.26) is therefore limited to low fields. Microwaves and visible light, which are used to probe the states of conduction electrons, have wavelengths much greater than the lattice constants. The Bloch electron should then respond to the electromagnetic fields carried by the radiation, as represented by Equation (10.32). If radiations, such as X rays and γ rays, whose wavelengths are comparable to or smaller than the lattice constants, are applied to a solid, the picture of the interaction between the Bloch electron (spread over several lattice units) and the radiation breaks down. Rather, the picture of the interaction between the free electron and the radiation should prevail; this picture is routinely used for theory of the photoelectric effect and the Compton scattering. In other words, Equation (10.26) is valid for the fields varying slowly over the lattice constant.

Earlier in Section 10.1, we pointed out the difficulties of Equation (10.6): $\hbar\dot{\mathbf{k}} = q(\mathbf{E} + \mathbf{v} \times \mathbf{B})$ in dealing with inhomogeneous or anisotropic systems. We complete our discussion of the comparison between Equation (10.6) and Equation (10.26) here. We show in particular that "electrons" and "holes" can be introduced naturally in consideration of the Fermi surface and Equation (10.26). As we shall see in the later chapters, correct knowledge of Bloch electron dynamics is essential in solid-state physics.

In the presence of a static magnetic field, a classical electron spirals around the field, keeping its kinetic energy unchanged. The motion of a Bloch electron in a metal is more complicated. But *its kinetic energy is conserved*. This is supported by the experimental fact that no joule heating is detected in any metals under slowly applied static magnetic fields. (We caution here that if the applied magnetic field is not changed slowly, some electric fields are necessarily generated, thereby causing a joule heating.)

The *energy conservation law* under a static magnetic field can be derived as follows. From Equation (10.15) with $\mathbf{E} = \Phi = 0$, we obtain

$$\frac{d}{dt} H_0(\mathbf{p} - q\mathbf{A}) = \sum_j \frac{\partial H_0}{\partial p_j} \frac{df_j}{dt} = \mathbf{v} \cdot q(\mathbf{v} \times \mathbf{B}) = 0 \tag{10.32}$$

[using Equations (10.17) and (10.21)]. If we choose the Cartesian coordinates along the principal axes of curvature, we obtain $df_i/dt = m_j(dv_j/dt)$. Hence we obtain

$$\sum_j \frac{\partial H_0}{\partial p_j} \frac{df_j}{dt} = \frac{d}{dt} \left(\sum_j \frac{1}{2} m_j v_j^2 \right) = 0. \tag{10.33}$$

We can then write the kinetic energy H_0 in the form:

$$H_0 = \sum_j \frac{1}{2} m_j v_j^2, \tag{10.34}$$

which is reduced to the familiar expression $E = mv^2/2$ in the free-electron limit. Equation (10.33) can now be integrated to yield

$$\boxed{H_0 \text{ (energy) } = \sum_j \frac{1}{2} m_j v_j^2 = \text{constant.}} \tag{10.35}$$

This conservation law cannot be derived from Equation (10.6), as the energy-momentum relation is not incorporated in this equation.

The principal masses $\{m_j\}$ defined by Equation (10.25) depend on $\mathbf{p} = \hbar\mathbf{k}$ and $\epsilon_n(\mathbf{p})$. No m_j can be zero. Otherwise, according to Equation (10.26), the acceleration becomes ∞ in the j-direction in the presence of the Lorentz force. This means that the Fermi surface cannot have cusps, as observed in all known experiments. This result is significant since the NFEM Fermi surface for a multivalent metal (see Section 9.1 and Figure 9.3) contains cusps; these cusps must be smoothed out.

Although Equation (10.26) is equivalent to Equation (10.29), $\sum_j m_{ij}(dv_j/dt) = q(\mathbf{E} + \mathbf{v} \times \mathbf{B})_i$, the former is much simpler to use, and it can be interpreted in terms of the principal masses. This is somewhat similar to the situation in which Euler's equations of motion for a rigid body are set up in the body's frame of coordinates and are expressed in terms of principal moments of inertia.

In our earlier discussion of the Hall effect, we saw that in some metals there are "holes" and/or "electrons." Their existence is closely connected with the actual Fermi surface. To illustrate this feature let us consider two cases:

1. Case of the "electron": the constant-energy surface is a sphere whose outward normal corresponds to the direction of increasing energy. The principal masses at any surface point are all equal to m_A, and they are positive according to Equation (10.26):

$$m_1 = m_2 = m_3 = m_A^* > 0. \tag{10.36}$$

2. Case of the "hole": the constant-energy sphere whose *inward* normal corresponds to the direction of increasing energy. The principal masses are all equal and negative:

$$m_1 = m_2 = m_3 = m_B^* < 0. \tag{10.37}$$

If we apply a constant magnetic field $\mathbf{B}$ in the positive x_3-direction, we observe from

$$m^* d\mathbf{v}/dt = (-e)(\mathbf{v} \times \mathbf{B}), \quad m^* = m_A^*, m_B^*, \tag{10.38}$$

that the electron moves counterclockwise (clockwise) in the x_1x_2-plane for case 1 (case 2) (see Figure 10.2).

The concept of a negative mass is a little strange. It is convenient to introduce a new interpretation as in the case in which Dirac discussed a positron in his prediction of this particle. In the Newtonian equations (10.23), the effective mass m^* and the electron charge $-e$ appear on the opposite sides. In the new interpretation, we say that the "hole" has a positive mass $|m^*| = -m^*$, a positive charge $+e$, and a positive energy. In this new picture, the linear momentum $|m^*|\mathbf{v}$ has the same direction as the velocity $\mathbf{v}$, and the kinetic energy $T = |m^*|v^2/2$ is positive. Hence, the dynamics of "electrons" and "holes" can be discussed in a parallel manner. Only we must remember that the "hole" has charge $+e$. We introduced "electrons" and "holes" by using the special constant-energy surfaces (spheres). To apply these concepts to a metal, we merely extend our theory to *local* Fermi surfaces. We discuss a few examples.

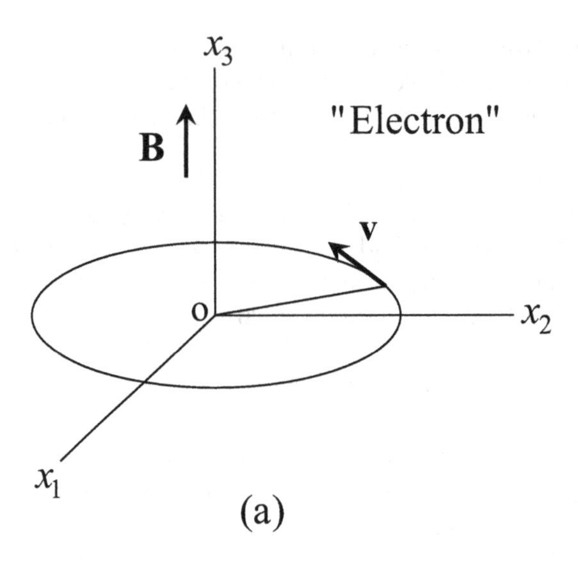

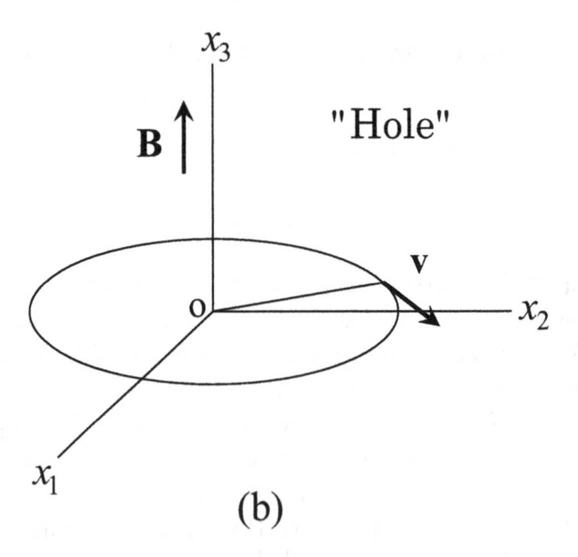

Figure 10.2: (a) With the magnetic field in the x_3-direction, a particle with mass $m_A > 0$ and charge $-e$ moves counterclockwise. Note that the Lorentz force $(-e)\mathbf{v} \times \mathbf{B}$ acts as the centripetal force. (b) A particle with mass $m_B < 0$ and charge $-e$ moves clockwise.

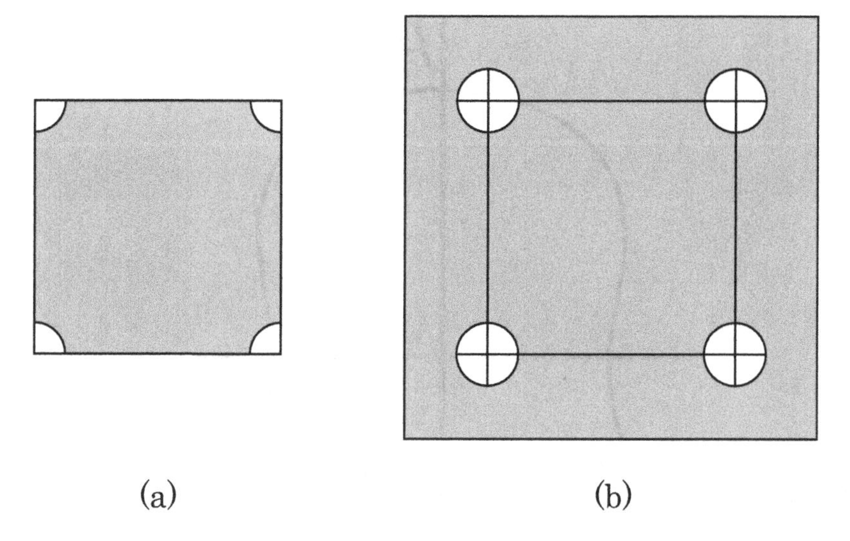

Figure 10.3: (a) Vacant states at the corners of an almost filled band in the reduced zone scheme. (b) The same states in the extended zone scheme.

Let us take a square lattice. According to Harrison's construction of a NFEM Fermi surface, the first Brillouin zone for a divalent metal may be filled with electrons everywhere except near the four corners as shown in Figure 10.3 (a). A possible electron orbit in the presence of a magnetic field **B** may be a circle in the extended-zone scheme shown in Figure 10.3 (b). By inspection the curvature $\kappa \equiv P^{-1}$ is positive, and therefore, the principal mass m^* is negative according to Equation (10.31). Applying Equation (10.26), we then observe that the Bloch electron moves like a "hole," that is, clockwise in the xy-plane if **B** is directed in the positive z-direction. The corresponding orbit in the position (or r-) space is a circle directed clockwise, which is in accord with the fact that the centripetal force must act for a circular motion. If we adopt Equation (10.6) with $q = -e$, then the k-vector changes clockwise in the $k_x k_y$-plane. To obtain a circular motion with the centripetal force, we must still assume that this hypothetical electron carries a negative mass. In other words, Equation (10.26) gives a complete dynamics, whereas Equation (10.6) does not. In particular, the known experimental fact that "holes" exist in some metals cannot be explained from Equation (10.6) alone.

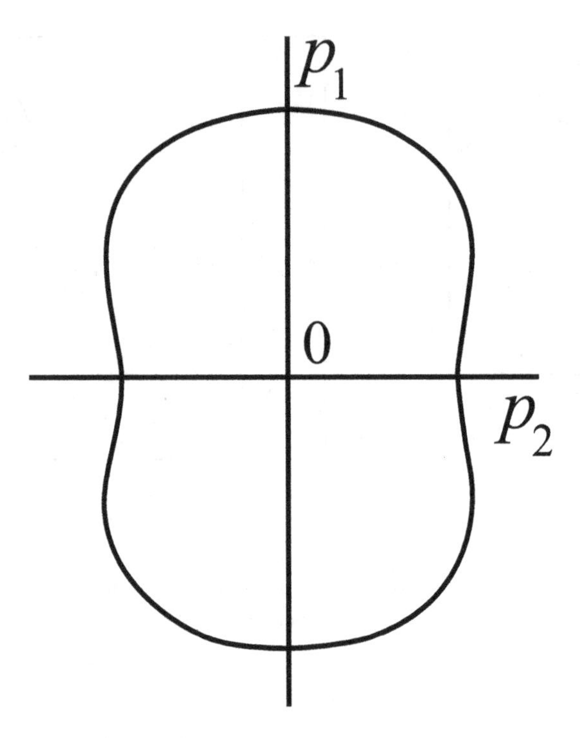

Figure 10.4: An oval of Cassini. The electron cannot travel on a closed p-curve that contains positive and negative curvatures.

Significantly different predictions arise when we consider a *closed Fermi curve* (an oval of Cassini) represented by

$$(p_1^2 + p_2^2)^2 + a^4 - 2a^2(p_1^2 - p_2^2) = b^2, \quad 2^{1/2}a > b \equiv (2m\epsilon_F)^{1/2} > a. \quad (10.39)$$

As shown in Figure 10.4, this curve contains arcs of negative curvatures where "electrons" with energies higher than the Fermi energy ϵ_F can move *and* arcs of positive curvatures where "holes" with energies less than ϵ_F can move. Since the magnetic field cannot supply energy, this is not a physically executable orbit. More generally, if any constant-energy curve contains arcs of curvatures of different signs, then the Bloch electron cannot travel along it. The prevailing equation (10.6) does not inhibit such motion.

One of the most important ways of studying the Fermi surface is to measure and analyze de Haas–van Alphen (dHvA) oscillations using Onsager's formula (12.2). As stated earlier, a physically executable Bloch electron k-orbit must consist of arcs of curvatures of the same sign. Onsager's formula may be applied only to those k-orbits.

In some solid-state physics texts, the "holes" are pictured as the "vacancies" that are created in otherwise filled electron sites in the ordinary r-space. This is not necessarily in agreement with the definition of the "holes" defined in terms of the curvatures of the Fermi surface in the k-space. We define the "hole" as a charged particle that circulates clockwise viewed from the tip of the applied magnetic field. The electric conduction occurs in the copper plane (CuO_2) in the cuprate superconductors. The number of electrons that nominally reside near the oxygen (O) sites [or the copper (Cu) sites] can be changed by the doping (substitution of atoms) at the neighboring planes. The change can cause the curvature inversion of the Fermi surface, generating "electrons" and/or "holes." This phenomenon can be observed by the Hall effect measurements. Yet no change other than the electron density change occurs in the copper plane (the r-space). In summary, "holes" are defined using the k-space, where the dispersion relation and the principal masses are defined, and are not defined in the position space alone.

Earlier in the discussion of the K-P model in Section 7.2, we saw that the effective mass m^* is positive at the lower edge of a band and it is negative at the upper edge. This behavior was illustrated in Figure 7.4. Note that in this energy (ϵ)-momentum (p) graph, the sign of the effective mass defined by $1/m^* = \partial^2\epsilon/\partial p^2 = \hbar^{-2}\partial^2\epsilon/\partial k^2$ can be recognized by inspection by examining the curvature of the ϵ-p curve. There is an *inflection point* along the curve, where the second derivative vanishes, and the effective mass m^* goes to ∞. Such singularity in the effective mass can occur in all directions (x_1, x_2, x_3). If the Fermi energy ϵ_F of a system of electrons moving in 3D is near such an inflection point, then the system shows a *heavy fermion behavior*: (1) the effective mass can be hundreds times the free electron mass, and (2) the density of states $\mathcal{N}(\epsilon_F)$ becomes exceptionally high since the gradient (magnitude) $|\nabla\epsilon(p)|$ vanishes in formula (9.16):

$$\mathcal{N}(\epsilon_F) = \frac{2}{(2\pi\hbar)^3}\int dS \frac{1}{|\nabla\epsilon(p)|}. \tag{10.40}$$

This phenomenon (*heavy fermion behavior*) has been observed in $CeCu_2Si_2$, $CeCu_6$, UPt_3, and others. The study of the heavy fermion systems, some of which are superconductors, is one of the current hot topics [12]. The cuprate superconductor $La_{2-x}Sr_xCuO_4$ has layered copper planes (CuO_2) in which the electric current runs. By doping Sr, the electron density in the copper plane and the Fermi energy can be changed in a controlled manner. We shall discuss this topic in book 2 [13].

Part III

Applications
Fermionic Systems (Electrons)

A selection of topics are discussed in Part III, Chapters 11 through 15. They are de Haas–van Alphen effect, magnetoresistance, cyclotron resonance, thermopower, and infrared Hall effect.

Chapter 11

De Haas–Van Alphen Oscillations

The de Haas–van Alphen oscillations in susceptibility are often analyzed, using Onsager's formula, which is derived. The statistical mechanical theory of the oscillations for the quasifree electron is also discussed in this chapter.

11.1 Onsager's Formula

A metal is a system in which the conduction electrons, either "electrons" or "holes," move without much resistance. "Electrons" ("holes") are fermions having negative (positive) charge $q = -(+)e$ and positive effective masses m^*, which respond to the Lorentz force $\mathbf{F} = q(\mathbf{E} + \mathbf{v} \times \mathbf{B})$. These conduction electrons are generated at finite temperatures in a metal, depending on the curvature sign of the metal's Fermi surface. Their existence can be checked by the linear heat capacity at the lowest temperatures. As we saw in Chapter 9, each metal has a Fermi surface, often quite complicated.

The most frequently used means to probe the Fermi surface is to observe de Haas–van Alphen (dHvA) oscillations and analyze the data using Onsager's formula, Equation (11.2). A magnetic field of the order 1 tesla (T) $= 10^4$ gauss is applied in a special lattice direction. The experiments are normally carried out using pure samples at the liquid helium temperatures to reduce the impurity and phonon scatterings. The carrier charge sign is not determined in this experimental probe.

The *magnetic susceptibility* χ is defined by

$$I = \chi B, \tag{11.1}$$

where I is the *magnetization* (magnetic moment per unit volume) and B the applied magnetic field (magnitude). When the experiments are done on pure samples and at very low temperatures, the susceptibility χ in some metals is found to exhibit oscillations with varying magnetic field strength B. This phenomenon was first discovered in 1930 by de Haas and van Alphen [1] in their study of bismuth (Bi), and it is called *de Haas–van Alphen effect* (*oscillations*). As we shall show, these oscillations have a quantum mechanical origin. Currently the analyses of the dHvA oscillations are done routinely in terms of Onsager's formula [Equation (11.2)]. According to Onsager's theory [2], the nth maximum (counted from $1/B = 0$) occurs for a field B given by the relation

$$\boxed{n + \gamma = \frac{1}{2\pi\hbar e}\frac{A}{B} \equiv \frac{1}{(2\pi\hbar)^2}\Phi_0\frac{A}{B},} \tag{11.2}$$

where A is any extremal area of intersection between the Fermi surface and the family of planes $\mathbf{B}\cdot\mathbf{p} \equiv \mathbf{B}\cdot(\hbar\mathbf{k}) =$ constant, and γ is a *phase* (number) less than unity. The constant

$$\Phi_0 \equiv \frac{2\pi\hbar}{e} \equiv \frac{h}{e} = 4.135 \times 10^{-1} \text{ gauss cm}^2 \tag{11.3}$$

is called the electron *flux quantum*.

As an example, consider an ellipsoidal Fermi surface, shown in Figure 11.1 (a):

$$\epsilon_F = \frac{p_1^2}{2m_1} + \frac{p_2^2}{2m_2} + \frac{p_3^2}{2m_3}, \qquad m_1, m_2, m_3 > 0. \tag{11.4}$$

The subscript F on ϵ will be omitted hereafter in this section. Assume that the field $\mathbf{B}$ is applied along the p_3-axis. All the intersections are *ellipses* represented by

$$\epsilon - \frac{p_3^2}{2m_3} = \frac{p_1^2}{2m_1} + \frac{p_2^2}{2m_2} \qquad (p_3 = \text{constant}). \tag{11.5}$$

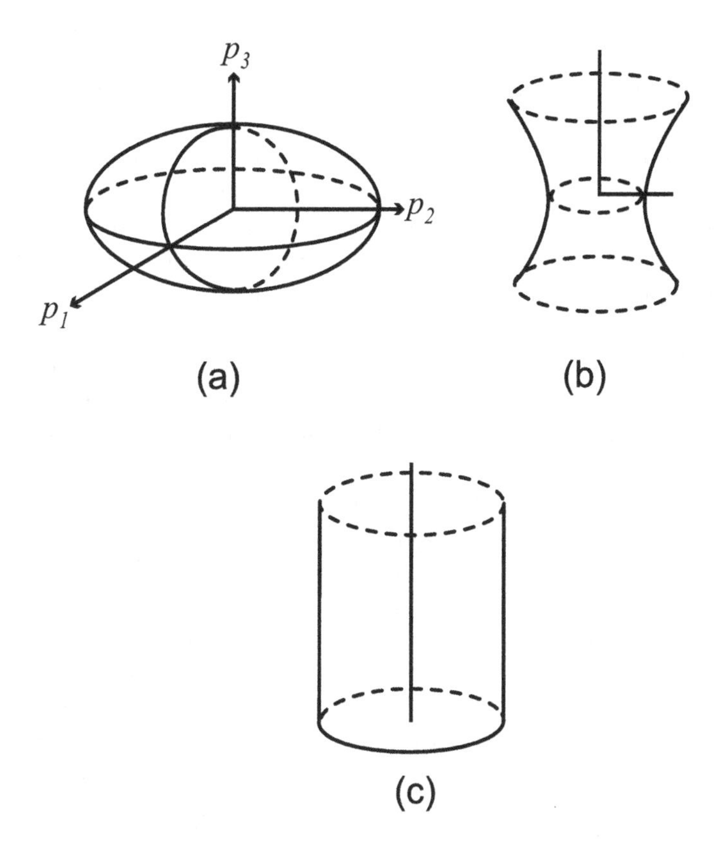

Figure 11.1: Fermi surfaces: (a) ellipsoid, (b) hyperboloid (neck), (c) cylinder.

The maximum area of the intersection occurs at $p_3 = 0$, "belly" [see Figure 11.1 (a)], and its area A is

$$A = \pi(2m_1\epsilon)^{1/2}(2m_2\epsilon)^{1/2} = 2\pi(m_1m_2)^{1/2}\epsilon. \tag{11.6}$$

Using Equation (11.2) and solving for ϵ, we obtain

$$\epsilon = eB(m_1m_2)^{-1/2}(n+\gamma)\hbar, \tag{11.7}$$

which indicates that the energy ϵ is quantized as the energy of the simple harmonic oscillator with the angular frequency

$$\omega_0 \equiv \frac{eB}{(m_1m_2)^{1/2}}. \tag{11.8}$$

As a second example, we take a *hyperboloidal* Fermi surface that can be represented by Equation (11.4) with m_1, $m_2 > 0$ and $m_3 < 0$. Assume the same orientation of $\mathbf{B}$. Equations (11.5) and (11.6) then hold, where the area A represents the minimal area of the intersection at $p_3 = 0$, "neck" [see Figure 11.1 (b)]. As a third example, assume that $m_3 = \infty$ in Equation (11.4), which represents a Fermi *cylinder* [see Figure 11.1 (c)]. In this case the area A is given by Equation (11.6) for the same orientation of $\mathbf{B}$.

All three geometrical shapes are discussed by Onsager in his correspondence [2]. At the time of his writing in 1952, only the ellipsoidal Fermi surface was known in experiments. Today we know that all three cases occur in reality. When tested by experiments, the agreements between theory and experiment are excellent. The cases of ellipsoidal and hyperboloidal surfaces were found in noble metals Cu, Ag, and Au. The dHvA oscillations in silver (Ag) are shown in Figure 11.2, where the susceptibility χ is plotted against B^{-1} in arbitrary units, after Schönberg and Gold [3]. The magnetic field is along a $\langle 111 \rangle$ direction. The two distinct periods are due to the "neck" and "belly" orbits indicated, the high-frequency oscillations coming from the larger belly orbit. By counting the number of high-frequency periods in a single low-frequency period, e.g., between the two arrows, we can deduce directly that $A_{111}(\text{belly})/A_{111}(\text{neck}) = 51$, which is most remarkable.

Onsager's derivation of Equation (11.2) in his original paper [2] is quite illuminating. Let us follow his arguments. For any closed k-orbit, there should be a closed orbit in the position space, called a closed r-orbit. The periodic component of the motion, which involves the components of $\mathbf{p} \equiv \hbar\mathbf{k}$ and $\mathbf{r}$ perpendicular to $\mathbf{B}$, is quantized. We apply the *Bohr–Sommerfeld quantization rule*

$$\oint \mathbf{p} \cdot d\mathbf{r} = (n + \gamma) 2\pi\hbar \tag{11.9}$$

to the r-orbit. The magnetic moment $\boldsymbol{\mu}$ is proportional to the angular momentum $\mathbf{j}$:

$$\boldsymbol{\mu} = \alpha\, \mathbf{j}, \qquad \alpha = \text{constant}. \tag{11.10}$$

The cross section Ω of the r-orbit is determined such that the enclosed magnetic flux Φ, given by $B\Omega$, equals $(n + \gamma)$ times the flux quantum $\Phi_0 \equiv h/e$:

$$\boxed{\Phi = (n + \gamma)\Phi_0 = B\Omega.} \tag{11.11}$$

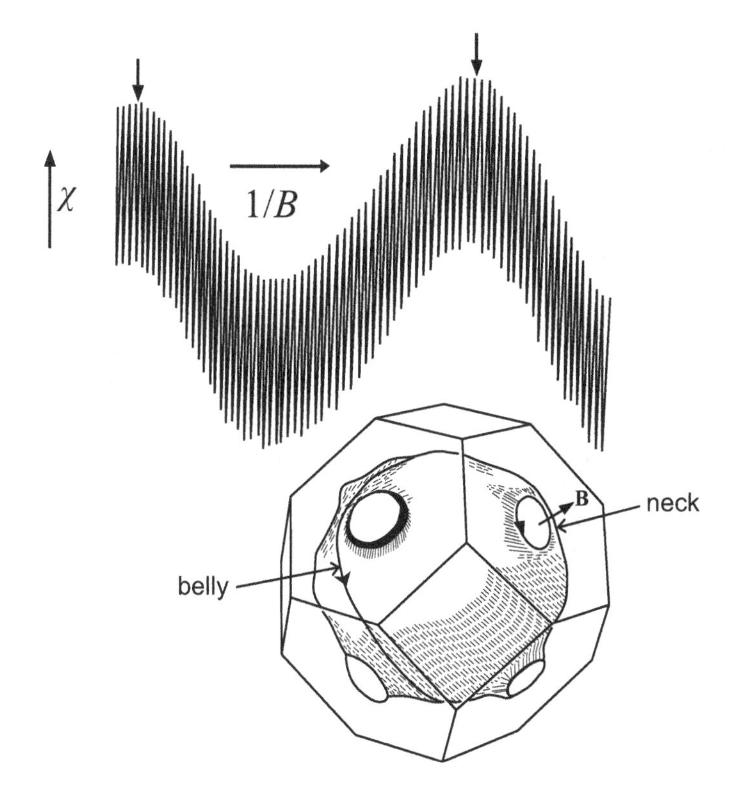

Figure 11.2: The dHvA oscillations in silver with the magnetic field along a $\langle 111 \rangle$ direction, after Schönberg and Gold [3]. The two distinct periods are due to the neck and belly orbits indicated in the inset.

This is Onsager's *magnetic flux quantization.*

For a free electron the closed circular path in the k-space perpendicular to the field becomes a similar path in the r-space, turned through a right angle, and with the linear dimension changed in the ratio $(eB)^{-1}$ (see Problem 11.1.1). This may hold for a nearly circular closed orbit. If we assume this relation for the Bloch electron, the area enclosed by the closed k-orbit, A, is proportional to that enclosed by the closed r-orbit, Ω:

$$A = (eB)^2 \Omega. \tag{11.12}$$

Combining the last two equations, we obtain Equation (11.2). QED.

The most remarkable argument advanced by Onsager is that an electron in a closed r-orbit may move, keeping a finite number n of flux quanta, each carrying $\Phi_0 \equiv h/e$, within the orbit. This comes from the physical principle that the magnetic field $\mathbf{B}$ does not work on the electron and, therefore, the field does not change the electron's kinetic energy. This property should hold for any charged particle. The flux quantization for the Cooper pair in a superconductor was observed in 1961 by Deaver and Fairbank [4] and Doll and Näbauer [5]. Because the Cooper pair has charge (magnitude) $2e$, the observed flux quanum is found to be $h/2e$, that is, half the electron flux quantum Φ_0 defined in Equation (11.3). The phase γ in Equation (11.9) can be set equal to 1/2. This can be deduced by taking the case of a free electron for which quantum calculations are carried out exactly. The quantum number n can arbitrarily be large. Hence, Onsager's formula can be applied for *any* strength of field (Problem 11.1.1).

However, Equation (11.2) turns out to contain a limitation. The curvatures along the closed k-orbit must either be entirely positive or negative. The k-orbit cannot have a mixture of a positive-curvature section and a negative-curvature section as explained in Section 10.2.

Problem 11.1.1. Consider a free electron having mass m and charge q, subject to a constant magnetic field $\mathbf{B}$.

a. Write down Newton's equation of motion.
b. Show that the magnetic force $q\mathbf{v}\times\mathbf{B}$ does not work on the electron; that is, the kinetic energy $E \equiv mv^2/2$ does not change with time.
c. Show that the component of $\mathbf{v}$ parallel to $\mathbf{B}$ is a constant.
d. Show that the electron spirals about the field $\mathbf{B}$ with the angular frequency $\omega = eB/m$.
e. Show that the orbit projected on a plane perpendicular to the field $\mathbf{B}$ is a circle of radius $R = v_\perp/\omega = mv_\perp/eB$, where $v_\perp$ represents the speed of the circular motion. Find the maximum radius R_m.
f. Define the *kinetic momentum* $\mathbf{\Pi} = m\mathbf{v}$ and express the energy ϵ in terms of Π_j.
g. Choose the $x_3(=z)$-axis along $\mathbf{B}$. Show that the curve represented by $E(p_1, p_2, 0) = \epsilon$ is a circle of radius $(2me)^{1/2} = \Pi$.
h. Show that the areas of the circles obtained in parts e and g differ by the factor $(eB)^2$.

11.2 Statistical Mechanical Calculations: 3D

Susceptibility is an equilibrium property and, therefore, can be calculated by using the standard statistical mechanics. Here, we demonstrate Onsager's formula (11.2) using a free-electron model.

The *free energy* F is, from Equation (5.34),

$$F = N\mu - 2k_BT \int_0^\infty dE \, \frac{dW}{dE} \ln\left[1 + e^{(\mu - E)/(k_BT)}\right]. \tag{11.13}$$

The oscillatory statistical weight $W_{\rm osc}$ is, from Equation (5.43),

$$W_{\rm osc} = A\frac{(\hbar\omega_c)^{3/2}}{\sqrt{2}\pi^{3/2}} \sum_{\nu=1}^{\infty} \frac{(-1)^\nu}{\nu^{3/2}} \sin\left(\frac{2\pi\nu E}{\hbar\omega_c} - \frac{\pi}{4}\right). \tag{11.14}$$

We note that $W_{\rm osc}$ oscillates with alternating signs. In fact the relevant E is of the order of the Fermi energy ϵ_F, which is much greater than the cyclotron frequency ω_c times the Planck constant $\hbar$. Hence, if there are many oscillations within the width of df/dE of the order k_BT, then the contribution to F must vanish. This condition is shown in Figure 11.3. Let us study this behavior in detail. Using Equation (5.44), we obtain

$$F = N\mu + 2\int_0^\infty dE \frac{df}{dE} \int_0^E dE' \, W_{\rm osc}(E'). \tag{11.15}$$

The critical temperature T_c below which the oscillations can be observed is

$$k_BT_c \sim \hbar\omega_c. \tag{11.16}$$

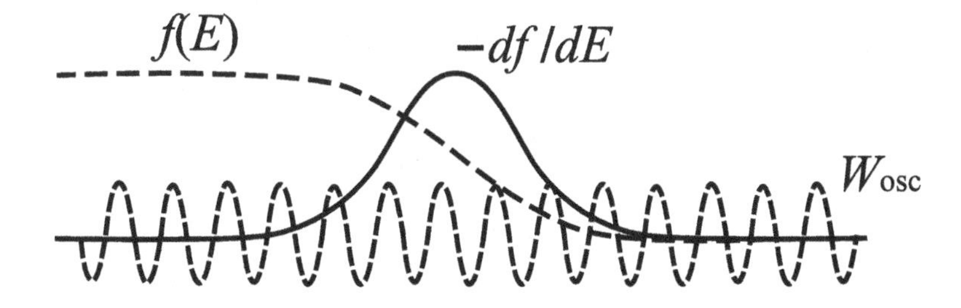

Figure 11.3: Numerous oscillations in $W_{\rm osc}$ within the width of $-df/dE$ cancel out the contribution to the free energy F.

Below this critical temperature ($T < T_c$), we cannot replace $-df/dE$ by $\delta(E-\mu)$ since the integrand varies violently. The integral in Equation (11.15)

$$\int_0^\infty dE \frac{df}{dE} \int_0^E dE' \, W_{\rm osc}(E') = \int_0^\infty dE \, W_{\rm osc}(E) f(E), \tag{11.17}$$

must be calculated with care. We introduce a new variable $\zeta = \beta E - \beta\mu$ and extend the lower limit to $-\infty$ ($\beta\mu \to \infty$)

$$\begin{aligned} \int_0^\infty dE \cdots \frac{1}{e^{\beta(E-\mu)}+1} &= \beta^{-1} \int_{-\mu\beta}^\infty d\zeta \cdots \frac{1}{e^\zeta + 1} \\ &\longrightarrow \beta^{-1} \int_{-\infty}^\infty d\zeta \cdots \frac{1}{e^\zeta + 1}. \end{aligned} \tag{11.18}$$

Using $\sin(A+B) = \sin A \cos B + \cos A \sin B$ and

$$\int_{-\infty}^\infty d\zeta \; e^{i\alpha\zeta} \frac{1}{e^\zeta + 1} = \frac{\pi}{i \sinh \pi\alpha}, \tag{11.19}$$

whose proof is given in Appendix C, we obtain from Equation (11.15)

$$F_{\rm osc} = A \frac{\sqrt{2}}{\sqrt{\pi}} (\hbar\omega_c)^{3/2} k_B T \sum_{\nu=1}^\infty \frac{(-1)^\nu}{\nu^{3/2}} \frac{\cos(2\pi\nu\epsilon_F/\hbar\omega_c - \pi/4)}{\sinh(2\pi^2 \nu k_B T/\hbar\omega_c)}. \tag{11.20}$$

Although Equation (11.20) contains an infinite sum with respect to ν just as the infinite sum in Equation (5.38), its summation character is quite different. Only the first term with $\nu = 1$ in the sum is important in practice because $[\sinh(2\pi^2 \nu k_B T/\hbar\omega_c)]^{-1} \ll 1$. Thus, we obtain

$$F_{\rm osc} = -A \frac{\sqrt{2}}{\sqrt{\pi}} (\hbar\omega_c)^{3/2} k_B T \frac{\cos(2\pi\epsilon_F/\hbar\omega_c - \pi/4)}{\sinh(2\pi^2 k_B T/\hbar\omega_c)}. \tag{11.21}$$

Using this equation we calculate the magnetization $I = -V^{-1}\partial F/\partial B$ and obtain

$$I_{\rm osc} = \frac{1}{\sqrt{2}} n \mu_B \frac{k_B T}{\epsilon_F} \left(\frac{\hbar\omega_c}{\epsilon_F}\right)^{1/2} \frac{\cos(2\pi\epsilon_F/\hbar\omega_c - \pi/4)}{\sinh(2\pi^2 k_B T/\hbar\omega_c)}. \tag{11.22}$$

The neglected terms are exponentially smaller than those in Equation (11.21) since $\exp(k_B T/\hbar\omega_c) \gg 1$. In the low field limit, the oscillation number in the range $k_B T$ becomes great, and hence, the contribution of the sinusoidal oscillations to the free energy must cancel out. This effect is represented by the factor $(\pi k_B T)[\sinh(2\pi^2 k_B T/\hbar\omega_c)]^{-1}$.

We define the *susceptibility* χ by

$$\chi = \frac{I}{B}. \tag{11.23}$$

Note that the magnetization I is not necessarily proportional to the field B. Using Equations (5.23), (5.49), (11.22), and (11.23), we obtain

$$\chi = \frac{1}{2}\frac{n\mu_B^2}{\epsilon_F}\left[3\left(\frac{m^*}{m}\right)^2 - 1 + \phi(T,B)\right], \tag{11.24}$$

$$\phi(T,B) = 2\sqrt{2}\frac{k_B T}{(\hbar\omega_c\epsilon_F)^{1/2}}\cos\left(2\pi\frac{\epsilon_F}{\hbar\omega_c} - \frac{\pi}{4}\right)e^{-2\pi^2 k_B T/\hbar\omega_c}. \tag{11.25}$$

Our calculations indicate that

1. the oscillation period is $\epsilon_F/\hbar\omega_c$. This result confirms Onsager's formula (11.2). In fact the maximum area of πp_F^2 occurs at $p_z = 0$. Hence $(2\pi\hbar)^{-2}(2\pi\hbar/e)(\pi p_F^2)/B = \epsilon_F/\hbar\omega_c$ if the quadratic dispersion relation $\epsilon = p^2/2m^*$ holds. We note that *all* electrons participate in the cyclotronic motion with the same frequency ω_c, and the signal is substantial.

2. The *envelope of the oscillations* exponentially decreases in B^{-1} as

$$\exp\left(\frac{-2\pi^2 k_B T}{\hbar\omega_c}\right) = \exp\left(-\frac{\delta}{B}\right), \qquad \delta \equiv \frac{2\pi^2 k_B T m^* \hbar}{e}. \tag{11.26}$$

   Thus if the "decay rate" δ in B^{-1} is measured carefully, the effective mass m^* may be obtained directly through

$$m^* = \frac{e\delta}{2\hbar\pi^2 k_B T}. \tag{11.27}$$

The calculations in this section were carried out by assuming a quasifree electron model. The actual physical condition in solids is more complicated. We cannot use the quasifree particle model alone to explain the experimental data, which will be discussed in Section 12.2.

11.3 Statistical Mechanical Calculations: 2D

The dHvA oscillations occur in 2D and 3D, but the 2D system is intrinsically paramagnetic since Landau's diamagnetism is absent, which is shown here.

Let us take a dilute system of quasifree electrons moving in a plane. Applying a magnetic field **B** perpendicular to the plane, each electron will be in the Landau state with the energy

$$E = \left(N_L + \frac{1}{2}\right)\hbar\omega_c, \qquad N_L = 0, 1, 2, \ldots . \tag{11.28}$$

We introduce kinetic momenta

$$\Pi_x = p_x + eA_x, \quad \Pi_y = p_y + eA_y. \tag{11.29}$$

The Hamiltonian $\mathcal{H}$ for the electron is then

$$\mathcal{H} = \frac{1}{2m^*}(\Pi_x^2 + \Pi_y^2) \equiv \frac{1}{2m^*}\Pi^2. \tag{11.30}$$

After simple calculations, we obtain

$$dx\, d\Pi_x\, dy\, d\Pi_y = dx\, dp_x\, dy\, dp_y. \tag{11.31}$$

We can then represent quantum states by the small quasi-phase space elements $dx\, d\Pi_x\, dy\, d\Pi_y$. The Hamiltonian $\mathcal{H}$ in Equation (11.30) does not depend on the position (x, y). Assuming large normalization lengths (L_1, L_2), we can represent the Landau states by the concentric shells having the statistical weight

$$2\pi\, \Pi\, \Delta\Pi \cdot L_1 L_2 (2\pi\hbar)^{-2} = \frac{eBA}{2\pi\hbar}, \tag{11.32}$$

with $A = L_1 L_2$ and $\hbar\omega_c = \Delta(\Pi^2/2m^*) = \Pi\Delta\Pi/m^*$ in the $\Pi_x\Pi_y$-space as shown in Figure 11.4. As the field B is raised the separation $\hbar\omega_c$ increases, and the quantum states are collected or bunched together. As a result of the bunching, the density of states, $\mathcal{N}(\epsilon)$, should change periodically since the Landau levels are equally spaced. The statistical weight W is the total number of states having energies less than

$$E = \left(N_L + \frac{1}{2}\right)\hbar\omega_c.$$

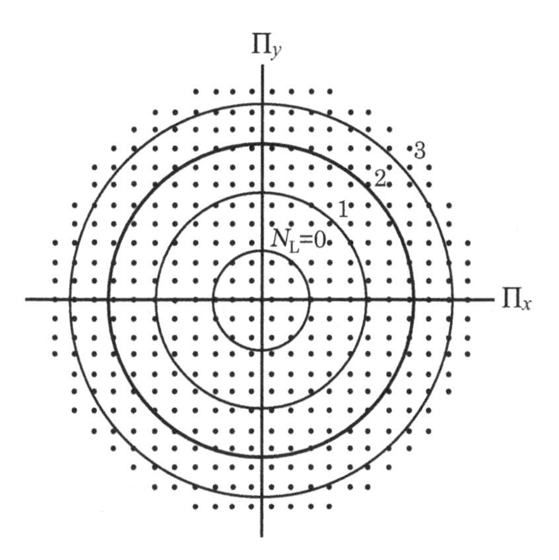

Figure 11.4: Quantization scheme for free electrons: without magnetic field (dots) and in a magnetic field (circles).

From Figure 11.4, this W is given by

$$W = \frac{L_1 L_2}{(2\pi\hbar)^2}\, 2\pi\, \Pi\, \Delta\Pi \cdot 2 \sum_{N_L}^{\infty} \Theta\left[E - \left(N_L + \frac{1}{2}\right)\hbar\omega_c\right], \tag{11.33}$$

where $\Theta(x)$ is the *Heaviside step function*:

$$\Theta(x) = \begin{cases} 1 & \text{if } x > 0 \\ 0 & \text{if } x < 0. \end{cases} \tag{11.34}$$

We introduce the dimensionless variable $\epsilon \equiv 2\pi E/\hbar\omega_c$, and rewrite W as

$$W(E) = C\,(\hbar\omega_c) 2 \cdot \sum_{N_L=0}^{\infty} \Theta\left[\epsilon - (2N_L + 1)\pi\right], \tag{11.35}$$

$$C = 2\pi m^* A (2\pi\hbar)^{-2}.$$

We assume a high Fermi degeneracy such that

$$\mu \simeq \epsilon_F \gg \hbar\omega_c. \tag{11.36}$$

The sum in Equation (11.35) can be computed by using Poisson's summation formula [6]:

$$\sum_{n=-\infty}^{\infty} f(2\pi n) = \frac{1}{2\pi} \sum_{m=-\infty}^{\infty} \int_{-\infty}^{\infty} d\tau f(\tau) e^{-i\omega\tau}. \tag{11.37}$$

After the mathematical steps detailed in Appendix B, we obtain [7]

$$W(E) = W_0 + W_{\text{osc}}, \tag{11.38}$$

$$W_0 = C\hbar\omega_c \left(\frac{\epsilon}{\pi}\right) = A\left(\frac{m^*}{\pi\hbar^2}\right) E, \tag{11.39}$$

$$W_{\text{osc}} = C\hbar\omega_c \frac{2}{\pi} \sum_{\nu=1}^{\infty} \frac{(-1)^\nu}{\nu} \sin\left(\frac{2\pi\nu E}{\hbar\omega_c}\right). \tag{11.40}$$

The B-independent term W_0 is the statistical weight for the system with no fields. The oscillatory term W_{osc} contains an infinite sum with respect to ν, but only the first term $\nu = 1$ is important in practice. This term W_{osc} can generate magnetic oscillations. There is no term proportional to B^2 generating tha Landau diamagnetism.

We calculate the free energy F in Equation (11.15) using the statistical weight W in Equations (11.38) through (11.40), and obtain

$$F = N\mu + A\frac{2m^*}{\pi\hbar^2}\epsilon_F + A\frac{2e}{\pi\hbar}Bk_BT \cdot \sum_{\nu=1} \frac{(-1)^\nu}{\nu} \frac{\cos(2\pi\nu\epsilon_F/\hbar\omega_c)}{\sinh(2\pi^2\nu k_B T m^*/\hbar e)}, \tag{11.41}$$

where we used the integration formulas in Equation (11.20). We took the low-temperature limit except for the oscillatory terms. The *magnetization* I, the total magnetic moment per unit area, can be obtained from

$$I = -\frac{1}{A}\frac{\partial F}{\partial B}. \tag{11.42}$$

Thus far, we did not consider the Pauli magnetization I_{Pauli} due to the electron spin [see Equation (5.18)]:

$$I_{\text{Pauli}} = \frac{2\mu_B^2}{A} N_0(\epsilon_F) = 2n\mu_B \frac{\mu_B B}{\epsilon_F}. \tag{11.43}$$

Using Equations (11.41) through (11.43), we obtain the total magnetization

$$\begin{aligned} I_{\text{tot}} &= I_{\text{Pauli}} + I \\ &= 2n\mu_B \frac{\mu_B B}{\epsilon_F} \left[1 - \left(\frac{\epsilon_F}{\mu_B B} \right) \frac{k_B T}{\epsilon_F} \left(\frac{m}{m^*} \right) \frac{\cos(2\pi\epsilon_F/\hbar\omega_c)}{\sinh(2\pi^2 m^* k_B T/\hbar e B)} \right]. \end{aligned} \tag{11.44}$$

In this calculation, we neglected the spurious contribution of the B-derivatives of the quantities inside the cosines $(2\pi\epsilon_F/\hbar\omega_c)$. This contribution is absent when we calculate the magnetization I through Equation (11.42) directly. The magnetic susceptibility χ is defined by the ratio

$$\chi = \frac{I}{B}. \tag{11.45}$$

Only the first oscillatory term, $\nu = 1$, is important and kept in Equation (11.44) since $\sinh(2\pi^2 m^* k_B T/\hbar e B) \gg 1$. The negative sign indicates a diamagnetic nature.

Figure 11.4 clearly shows that all electrons, not just those excited electrons near the Fermi surface, are subject to the magnetic field and all are in the Landau states. This is reflected by the fact that the Pauli magnetization I_{Pauli} is proportional to the electron density n, as seen in Equation (11.43). The oscillatory magnetization is also proportional to the density n. At a finite temperature T, $-df/dE$ has a width. In this E-range of the order of $k_B T$ many oscillations occur if the field B is lowered. Assuming this condition, we obtained Equation (11.44), and hence, this equation is valid for any finite T. At $T = 0$, the width vanishes and the oscillatory terms also vanishes.

When the system is subjected to an external electric field, all electrons will respond, and hence, the magnetoconductivity σ should be proportional to the electron density n. The oscillatory statistical weight generates a Shubnikov–de Haas (SdH) oscillation [8] with the factor smaller by the factor $k_B T/\epsilon_F$ in σ. Hence the dHvA and SdH oscillations should be similar.

Störmer et al. [7] measured the dHvA oscillations at 1.5 K in GaAs/AlGaAs, by stacking 4000 layers equivalent to the area of 240 cm^2. (Without stacking the signal is too small to observe.) Their data are shown in Figure 11.5. We see here that (1) the oscillation periods $\epsilon_F/\hbar\omega_c$ match, (2) the magnetization $(-I)$ rather than the susceptibility χ behaves more similarly to the diagonal

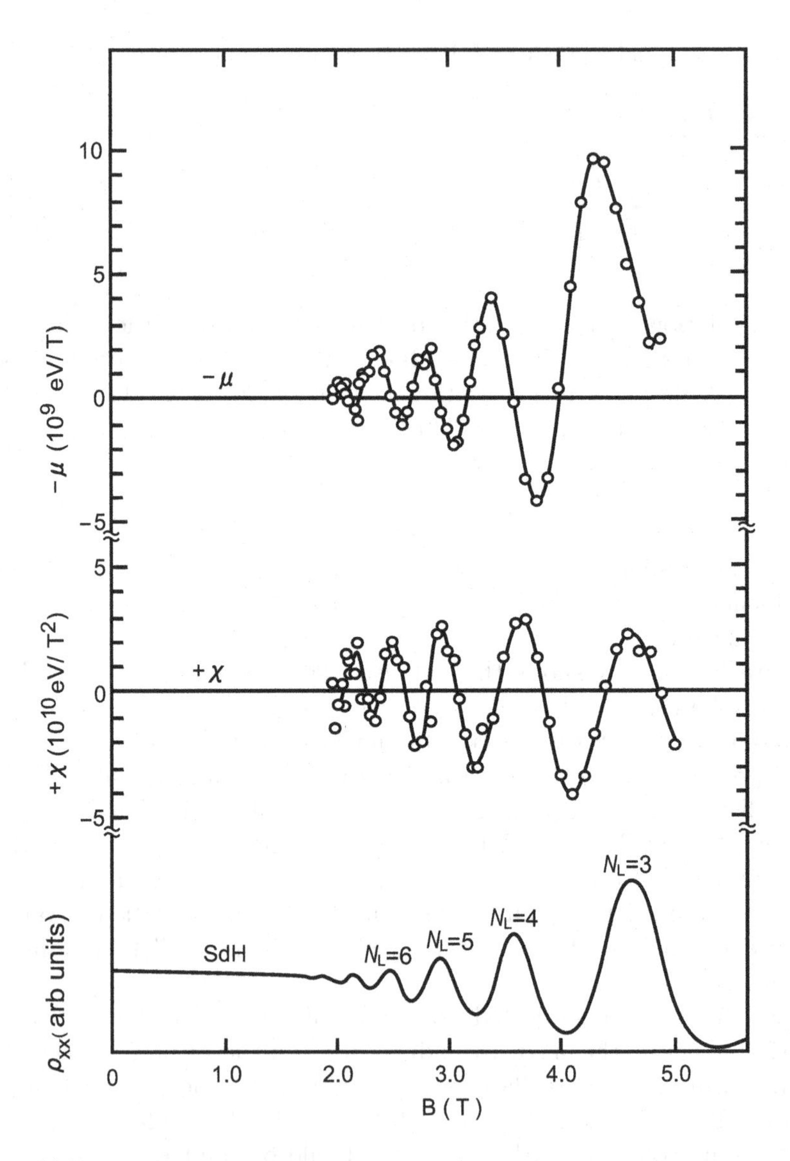

Figure 11.5: Experimental results on stack of 4000 layers of 2D electron systems equivalent to an area of 240 cm^2 after Störmer et al [7]. μ is the magnetic moment, and χ is the susceptibility. The trace denoted SdH are Shubnikov–de Haas data on a separate specimen of the same sample.

resistance ρ_{xx}, (3) the central line of the oscillation (background) is roughly independent of the field, and (4) the envelopes for $-I$ and ρ_{xx} are similar.

The feature (1) means that both oscillations arise from the same cause, the periodic oscillation of the statistical weight W. The feature (2) simply comes from the same field dependence (the B-independence) of the background (central line) of the oscillations. The density of states for a 2D quasifree electron system with no field is independent of the energy. This behavior (3) means that GaAs/AlGaAs is described adequately by this quasifree electron model. The feature (4) requires a further discussion. Our formula (11.44) indicates that the period of the oscillations, $\epsilon_F/\hbar\omega_c = m^*\epsilon_F/\hbar eB$, and the exponential decay rate, $\pi k_B T/\hbar\omega_c = \pi k_B T m^*/\hbar e$, of the envelope $[\sinh(2\pi^2 k_B T m^*/\hbar eB)]^{-1}$ are both controlled by the effective mass m^*. This feature is found not supported by the observed experiments. We shall discuss this point later in Chapter 12.

In conclusion, the 2D quasifree electron system is intrinsically paramagnetic, since there is no Landau diamagnetism. However, there are magnetic oscillations. The AlGaAs/GaAs/AlGaAs heterostructure is often used for the study of the quantum Hall effect (QHE). The parental 3D GaAs is diamagnetic, and hence, the magnetic behavior is greatly different in 2D and 3D.

11.4 Two-Dimensional Conductors

Layered materials like graphite (C) become 2D conductors. The conductivity measured is a few orders of magnitude smaller along the c-axis than perpendicular to it. This appears to contradict the prediction based on the naive application of the Bloch theorem. This puzzle may be solved as follows [10].

Suppose an electron jumps from one conducting layer to its neighbor. This generates a change in the charge states of the layers involved. If each layer is macroscopic in dimension, the charge state Q_n of the nth layer can change without limits: $Q_n = \ldots, -2, -1, 0, 1, 2, \ldots$ in units of the electron charge e. Because of unavoidable short circuits between layers due to the lattice imperfections, Q_n may not be large. If Q_n are distributed *at random* over all layers, the periodicity of the potential for the electron along the c-axis is lost. Therefore, the Bloch theorem based on the electric potential periodicity does not apply along the c-axis even though the lattice is perfectly periodic. There are no k-vectors along the c-axis. This means that the effective mass along the c-axis is infinity, so that the Fermi surface for the layered conductor is a right cylinder with its axis along the c-axis. In other words, the electron moves in the ab-plane.

The most direct way of verifying the 2D structure is to observe the orientation dependence of the cyclotron resonance peaks (see Section 13.2). The peak position (ω) in general follows Shockley's formula [9] [see Equation (13.9)],

$$\frac{\omega}{eB} = \left(\frac{m_2 m_3 \cos^2(\mu, x_1) + m_3 m_1 \cos^2(\mu, x_2) + m_1 m_2 \cos^2(\mu, x_3)}{m_1 m_2 m_3} \right)^{1/2}, \tag{11.46}$$

where (m_1, m_2, m_3) are effective masses along the (a, b, c) crystal axes and $\cos(\mu, x_j)$ is the direction cosine relative to the field $\mathbf{B}$ and the axis x_j. If the electron motion is plane-restricted, so that $m_3 \to \infty$, Equation (11.46) is reduced to the *cosine law* formula:

$$\omega = eB(m_1 m_2)^{-1/2} \cos\theta, \tag{11.47}$$

where θ is the angle between the field and the c-axis. A second and much easier way of verifying a 2D conduction is to measure the dHvA oscillations and analyze the orientation dependence of the dHvA frequency with the help of Onsager's formula (11.2):

$$\Delta \left[\frac{1}{B} \right] = \frac{2\pi e}{\hbar} \frac{1}{A}, \tag{11.48}$$

where A is the extremum intersectional area of the Fermi surface and the planes normal to the applied magnetic field $\mathbf{B}$. Wosnitza et al. [10], reported the first direct observation of the angular dependence ($\cos\theta$ law) of the dHvA oscillations in κ-$(ET)_2Cu(NCS)_2$ and α-$(ET)_2(NH_4)Hg(SCN)_4$, both layered organic superconductors, confirming a right cylindrical Fermi surface. Their data and theoretical curves are shown in Figure 11.6. Notice the excellent agreement between theory and experiment.

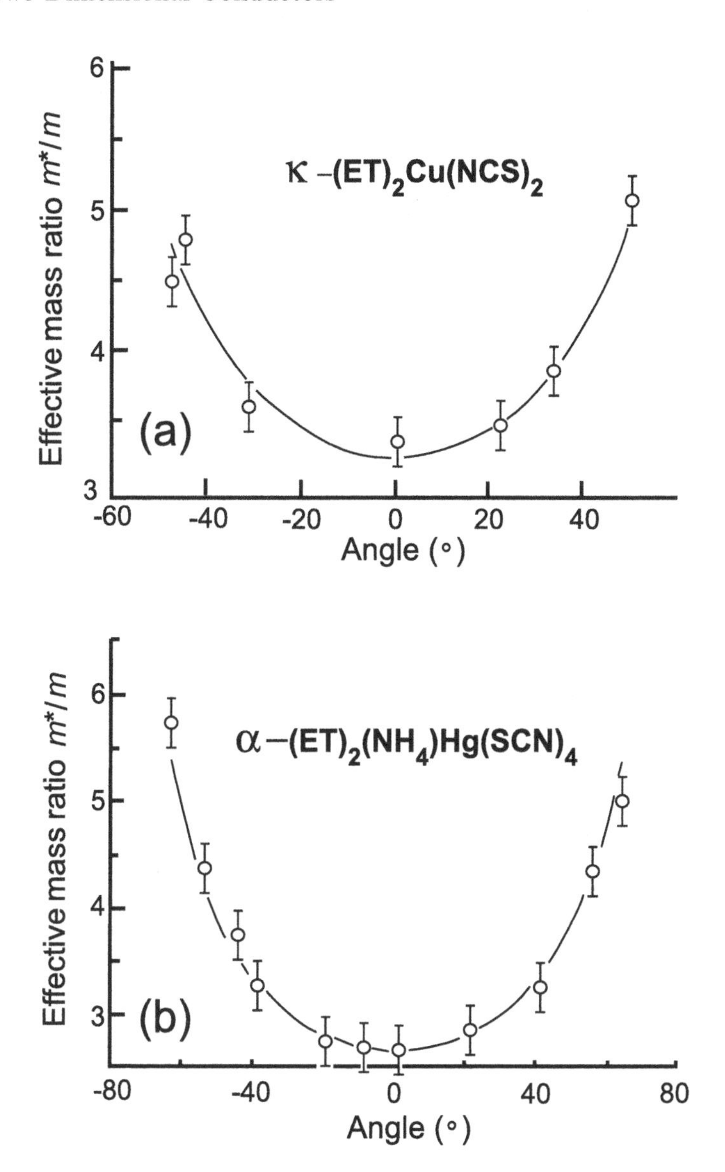

Figure 11.6: Angular dependence of the reduced effective mass in (a) κ-$(ET)_2Cu(NCS)_2$ and (b) α-$(ET)_2(NH_4)Hg(SCN)_4$. An angle of 0° means H is perpendicular to the conducting plane. The solid fits are obtained using Equations 11.47 and 11.48. After Wosnitza et al. [10].

Chapter 12

Magnetoresistance

Magnetoresistance (MR) in general is nonzero and anisotropic. A spectacular anisotropy observed in Cu is explained based on the nonspherical Fermi surface of this metal in this chapter. Magnetic oscillations found in the susceptibility also manifest themselves in magnetoresistance at low temperatures. A quantum theory is developed for the Shubnikov–de Haas oscillation for a 2D system. The period of the oscillations in B^{-1} is $\epsilon_F/\hbar\omega_c$, where ϵ_F is the Fermi energy and $\omega_c = eB/m^*$ is the cyclotron frequency. The envelope of the oscillations is proportional to $[\sinh(2\pi^2 M^* k_B T/\hbar e B)]^{-1}$, where M^* is the magnetotransport mass. Comparison between theory and experiment for heterojunction GaAs/AlGaAs *directly* gives the value $M^* \sim 0.30m$, which is 4.5 times heavier than the cyclotron mass $m^* \sim 0.067m$. The magnetotransport carriers are the composite fermions (c-fermions), each electron with two flux quanta. The present theory avoids the need of the Dingle temperature mysteriously arising from the "Landau level damping."

12.1 Introduction

Let us consider a rectangular metal sample shown in Figure 12.1. The current runs in the x-direction. The magnetic field $\mathbf{B}$ is applied in the z-direction. The *Hall voltage* V_H and the *Hall field* E_H are developed (positively or negatively) in the y-direction. The resistivity $\rho(B)$ along the x-direction may vary with the magnetic field magnitude. The *transverse magnetoresistance*

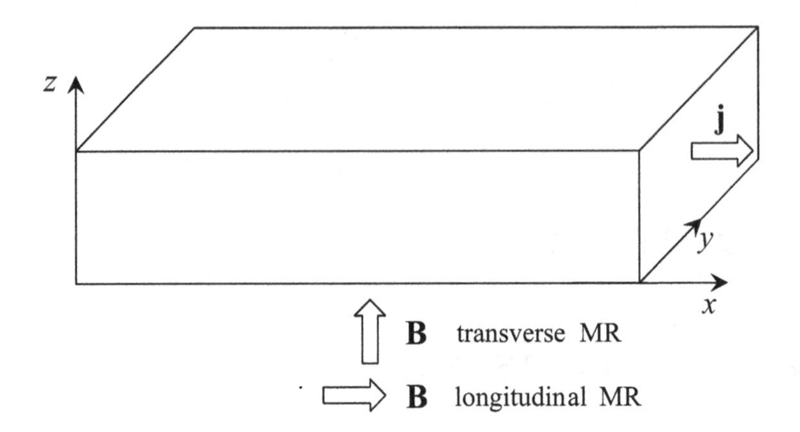

Figure 12.1: The current runs in the x-direction. When the B-field is applied perpendicular to (parallel with) the current direction, the transverse (longitudinal) magnetoresistance $\Delta\rho \equiv [\rho(B) - \rho_0]/\rho_0$ is generated.

$\Delta\rho$ is defined by

$$\Delta\rho \equiv \frac{\rho(B) - \rho_0}{\rho_0}, \qquad \rho_0 \equiv \rho(B = 0), \tag{12.1}$$

where ρ_0 is the resistivity without the field. Note that the MR is a dimensionless number. The magnetic field also can be applied along the sample length. The magnetoresistance defined in the form (12.1) is then called the *longitudinal magnetoresistance*.

Traditionally, the electric transport has been treated using kinetic theory or the Boltzmann equation method. In the presence of a static magnetic field the classical electron orbit is curved. Then, the basic *kinetic theoretical picture* in which the electron moves on a straight line hits an scatterer (impurity), changes direction, and moves on another straight line breaks down. Furthermore, the Boltzmann collision terms containing the scattering cross section cannot be written down.

Fortunately, quantum theory can save the situation. If the magnetic field is applied, the classical electron can continuously change from the straight-line motion at zero field to the spiral motion at a finite B. When the magnetic field is applied slowly, the energy of the electron does not change, but the spiral motion always acts to reduce the magnetic field energy. Hence, the total energy of the electron with its surrounding field is less than the electron energy plus the unperturbed field energy. The electron "dressed" with the field is stable against the breakup, and it is in a bound (negative energy) state. If an electric field is applied in the x-direction, the *dressed electron*, whose position is the *guiding center* of the circulation, preferentially jumps in the x-direction and generates a current. Thus, we can apply kinetic theory to the guiding center motion. We obtain an expression for the electrical conductivity σ:

$$\boxed{\sigma = \frac{e^2}{M^*} n \, \tau,} \tag{12.2}$$

where n is the density of the dressed electrons, e the charge, M^* the *magnetotransport* (effective) *mass* , and τ the relaxation time. The magnetotransport mass M^* is different from the cyclotron mass m^*.

Pippard in his seminal book, *Magnetoresistance in Metals* [1], argued that the magnetoresistance for the quasifree electron system vanishes;

$$\Delta\rho = 0, \tag{12.3}$$

after using the *relaxation time approximation* in the Boltzmann equation method. Magnetoresistance is finite in actual experimental condition. In fact, Equation (12.2) contains the magnetotransport mass M^*, which is distinct from the cyclotron mass m^*. This fact alone makes $\Delta\rho$ nonzero.

12.2 Anisotropic Magnetoresistance in Cu

If the Fermi surface is nonspherical, then the MR $\Delta\rho$ is anisotropic. In particular, Cu has a so-called "open" orbit in k-space as shown in Figure 12.2 (b). This open orbit contains positive and negative curvatures along the energy contour, and hence, it cannot be traveled by the physical electron as stated earlier in Section 10.3. Klauder and Kunzler [2] observed a striking angle-dependent MR as shown in Figure 12.3. The MR is more than 400 times the zero-field resistance in some directions.

The anisotropy may be interpreted as follows. We consider the narrow limits of the "necks," that is, *singular points*. There are eight (8) singular points in total, as seen from the figure. If the field $\mathbf{B}$ is in [001], there

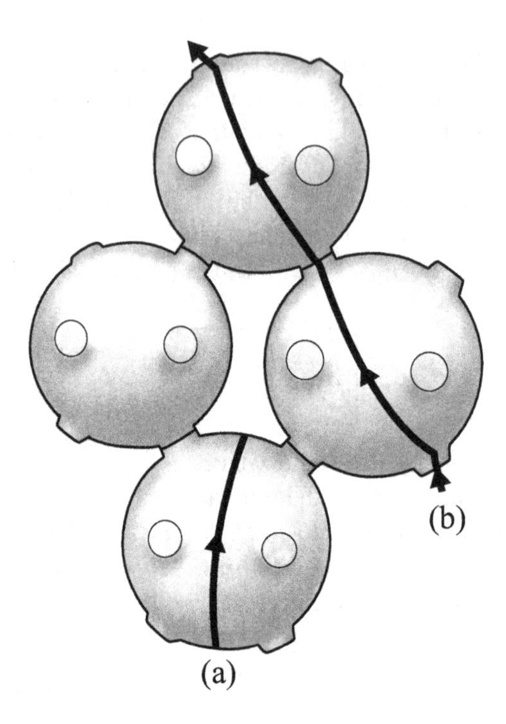

Figure 12.2: (a) A "closed" orbit a in k-space that can be traced by the electron. (b) An "open" orbit b that extends over two Brillouin zones and that cannot be traveled by the electron.

are one plane perpendicular to [001] containing four points and another containing four points. The same condition also holds when the field **B** is in [010]. This condition corresponds to the major minimal MR as seen in the data. Consider now the case where the field **B** is in [011]. There are three planes perpendicular to [011] that contain (2, 4, 2) singular points. This case corresponds to the second minimum in $\Delta\rho$. The two broad minima in the data correspond to the case where the field **B** is such that there are four planes perpendicular to **B**, each containing two singular points. There is a range of angle in which this condition holds. Hence, these minima are broad. In actuality each singular point is a finite neck. The situation, therefore, is more complicated. Our interpretation applies qualitatively.

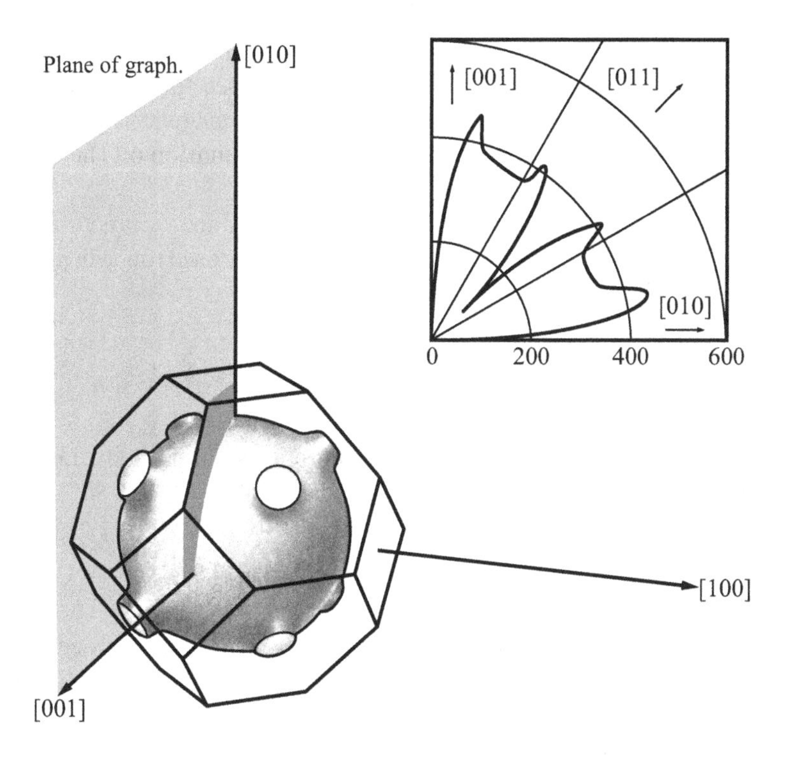

Figure 12.3: The striking anisotropy of the MR $\Delta\rho$ in Cu after Klauder and Kunzler [2]. The [001] and [010] directions of the copper crystal are shown, and the current flows in the [100] direction perpendicular to the graph. The magnetic field is in the plane of the graph. Its magnitude is fixed at 18 kilogauss, and its direction varied continuously from [001] to [010].

12.3 Shubnikov–De Haas Oscillations

Oscillations in magnetoresistance, similar to the dHvA oscillations in the magnetic susceptibility, were first observed by Shubnikov and de Haas in 1930 [3]. These oscillations are often called the *Shubnikov–de Haas* (SdH) *oscillations*. The susceptibility is an equilibrium property and can therefore be calculated by standard statistical mechanical methods. The MR is a nonequilibrium property, and its treatment requires a kinetic theory or the

Boltzmann equation, but the magnetic oscillations in both cases arise from the periodically varying density of states. We shall see that the observation of the oscillations gives a direct measurement of the magnetotransport mass M^*. The observation also gives the quantitative information on the cyclotron mass m^*.

Let us take a dilute system of electrons moving in a plane. Applying a magnetic field **B** perpendicular to the plane, each electron will be in the Landau state with the energy

$$E = \left(N_L + \frac{1}{2}\right)\hbar\omega_c, \quad \omega_c \equiv \frac{eB}{m^*}, \quad N_L = 0,\, 1,\, 2,\, \ldots, \tag{12.4}$$

where m^* is the *cyclotron mass*. The degeneracy of the Landau Level (LL) is given by Equation (4.32):

$$\frac{eBA}{2\pi\hbar}, \qquad A = \text{sample area}. \tag{12.5}$$

The weaker is the field, the more LLs, separated by $\hbar\omega_c$, are occupied by the electrons. In this Landau state the electron can be viewed as circulating around the guiding center. The radius of circulation $l \equiv (\hbar/eB)^{1/2}$ for the Landau groundstate is about 250 Å at a field 1.0 T (tesla). If we now apply a weak electric field **E** in the x-direction, then the guiding center preferentially jumps and generates a current.

Let us first consider the case with no magnetic field. We assume a uniform distribution of impurities with the density n_I. Solving the Boltzmann equation [see Equation (6.26)], we obtain the conductivity

$$\sigma = \frac{2e^2}{m^*(2\pi\hbar)^2}\int d^2p\,\frac{\epsilon}{\Gamma}\left(-\frac{df}{d\epsilon}\right), \qquad \epsilon = \frac{p^2}{2m^*}, \tag{12.6}$$

where Γ is the energy (ϵ)-dependent relaxation rate, Equation (6.23).

$$\Gamma(\epsilon) = n_I \int d\Omega\,\left(\frac{p}{m^*}\right) I(p,\theta)(1-\cos\theta), \tag{12.7}$$

with θ = scattering angle and $I(p,\theta)$ = scattering crosssection, and the Fermi distribution function

$$f(\epsilon) \equiv \left[e^{\beta(\epsilon-\mu)}+1\right]^{-1} \tag{12.8}$$

with $\beta \equiv (k_B T)^{-1}$ and μ = chemical potential is normalized such that

$$n = \frac{2}{(2\pi\hbar)^2} \int d^2p\, f(\epsilon), \tag{12.9}$$

where the factor 2 is due to the spin degeneracy. We introduce the density of states, $\mathcal{N}(\epsilon)$, such that

$$\frac{2}{(2\pi\hbar)^2} \int d^2p \cdots = \int d\epsilon\, \mathcal{N}(\epsilon) \cdots . \tag{12.10}$$

We can then rewrite Equation (12.6) as

$$\sigma = \frac{e^2}{m^*} \int_0^\infty d\epsilon\, \mathcal{N}(\epsilon) \frac{\epsilon}{\Gamma} \left(-\frac{df}{d\epsilon} \right). \tag{12.11}$$

The Fermi distribution function $f(\epsilon)$ drops steeply near $\epsilon = \mu$ at low temperatures:

$$k_B T \ll \epsilon_F. \tag{12.12}$$

The density of states, $\mathcal{N}(\epsilon)$, is a slowly varying function of the energy ϵ. For a 2D quasifree electron system, the density of states is independent of the energy ϵ. Then, the Dirac *delta-function replacement formula*

$$-\frac{df}{d\epsilon} = \delta(\epsilon - \mu) \tag{12.13}$$

can be used. Assuming this formula, using

$$\int_0^\infty d\epsilon\, \mathcal{N}(\epsilon)\epsilon \left(-\frac{df}{d\epsilon} \right) = \int_0^\infty d\epsilon\, \mathcal{N}(\epsilon) f(\epsilon), \tag{12.14}$$

and comparing Equations (12.6) and (12.11), we obtain

$$\gamma^{-1} = \int_0^\infty d\epsilon\, \mathcal{N}(\epsilon) \frac{1}{\Gamma(\epsilon)} f(\epsilon). \tag{12.15}$$

Note that the temperature dependence of the relaxation rate γ is introduced through the Fermi distribution function $f(\epsilon)$.

Next we consider the case with a magnetic field. We assume that a dressed electron is a fermion with magnetotransport mass M^* and charge

e. Applying kinetic theory to the dressed electrons, we obtain the standard formula (12.2): $\sigma = e^2 n\tau/M^*$. As discussed earlier, the dressed electrons move in all directions (isotropically) in the absence of the field.

We introduce *kinetic momenta*

$$\Pi_x = p_x + eA_x, \qquad \Pi_y \equiv p_y + eA_y. \tag{12.16}$$

The quasifree electron Hamiltonian H is

$$H = \frac{1}{2m^*}(\Pi_x^2 + \Pi_y^2) \equiv \frac{1}{2m^*}\Pi^2. \tag{12.17}$$

The variables $(\Pi_x,\, \Pi_y)$ are the same kinetic momenta introduced earlier in Equation (4.34). Only we are dealing with a 2D system here. After simple calculations, we obtain

$$dx\, d\Pi_x\, dy\, d\Pi_y \equiv dx\, dp_x\, dy\, dp_y. \tag{12.18}$$

We can represent the quantum states by quasi-phase space elements $dx\, d\Pi_x\, dy\, d\Pi_y$. The Hamiltonian H in Equation (12.17) does not depend on the position $(x,\, y)$. Assuming large normalization lengths (L_1, L_2), $A = L_1L_2$, we can then represent the Landau states by the concentric shells in the $\Pi_x\Pi_y$- space (see Figure 4.4), having the statistical weight

$$(2\pi)\Pi\Delta\Pi L_1L_2 \cdot (2\pi\hbar)^{-2} = A(2\pi\hbar)^{-1}\omega_c m^* = \frac{eAB}{2\pi\hbar} \tag{12.19}$$

with the energy separation $\hbar\omega_c = \Delta(\Pi^2/2m^*) = \Pi\Delta\Pi/m^*$. Equation (12.19) confirms that the LL degeneracy is $eBA/(2\pi\hbar)$, as stated in Equation (12.5).

Let us consider the motion of the field-dressed electrons (guiding center). We assume that the dressed electron is a fermion with magnetotransport mass M^* and charge e. The kinetic energy is represented by

$$\mathcal{H}_K = \frac{\Pi_x^2 + \Pi_y^2}{2M^*} \equiv \frac{\Pi^2}{2M^*}. \tag{12.20}$$

According to Onsager's flux quantization represented by Equation (11.11), the magnetic fluxes can be counted in units of $\Phi_0 = e/h$. The dressed electron is composed of an electron and two elementary fluxes (fluxons). A further explanation of the present model will be given later.

Let us introduce a *distribution function* $\varphi(\mathbf{\Pi}, t)$ in the $\Pi_x \Pi_y$- space normalized such that

$$\frac{2}{(2\pi\hbar)^2} \int d^2\Pi \ \varphi(\Pi_x, \ \Pi_y, \ t) = \frac{N}{A} = n. \tag{12.21}$$

The Boltzmann equation for a homogeneous stationary system is

$$e(\mathbf{E} + \mathbf{v} \times \mathbf{B}) \cdot \frac{\partial \varphi}{\partial \mathbf{\Pi}} = \int d\Omega \, \frac{\Pi}{M^*} n_I I(\Pi, \theta) \left[\varphi(\mathbf{\Pi}') - \varphi(\mathbf{\Pi})\right], \tag{12.22}$$

where θ is the angle of deflection, that is, the angle between the initial and final kinetic momenta ($\mathbf{\Pi}$, $\mathbf{\Pi}'$). In the actual experimental condition the magnetic force term can be neglected; detailed explanation is given in Section 12.4. Assuming this condition for now, we obtain the same Boltzmann equation for a field-free system [see Equation (6.8)]. Hence, we obtain the same conductivity formula (12.6) with m^* replaced by M^*, yielding Equation (12.2).

As the field B is raised, the separation $\hbar\omega_c$ becomes greater and the quantum states are bunched together. The density of states should contain an oscillatory part

$$\sin\left(\frac{2\pi\epsilon'}{\hbar\omega_c} + \phi_0\right), \qquad \epsilon' = \frac{\Pi'^2}{2m^*}, \tag{12.23}$$

where ϕ_0 is a phase. Since

$$\epsilon_F/\hbar\omega_c \gg 1 \qquad \text{(weak field)}, \tag{12.24}$$

the phase ϕ_0 will be dropped hereafter. Physically, the sinusoidal variations in Equation (12.23) arise as follows. From the Heisenberg uncertainty principle (phase space consideration) and the Pauli exclusion principle, the Fermi energy ϵ_F remains approximately constant as the field B varies. The density of states is high when ϵ_F matches the N_Lth level, while it is small when ϵ_F falls between neighboring LLs.

If the density of states, $\mathcal{N}(\epsilon)$, oscillates violently in the drop of the Fermi distribution function $f(\epsilon) \equiv (e^{\beta(\epsilon-\mu)}+1)^{-1}$, one cannot use the delta-function replacement formula. The use of Equation (12.13) is limited to the case in which the integrand is a smooth function near $\epsilon = \mu$. The width of $df/d\epsilon$ is of the order k_BT. The critical temperature T_c below which the oscillations can be observed is

$$k_BT_c \sim \hbar\omega_c. \tag{12.25}$$

Below the critical temperature $T < T_c$, we may proceed as follows. Let us consider the integral

$$I = \int_0^\infty d\epsilon\, f(\epsilon)\sin\left(\frac{2\pi\epsilon'}{\hbar\omega_c}\right), \quad \epsilon \equiv \frac{\Pi^2}{2M^*}. \tag{12.26}$$

For the temperatures satisfying $\beta\epsilon = \epsilon/k_BT \gg 1$, we can use the same mathematical steps as going from Equation (11.17) to Equation (11.20) and obtain

$$I = \pi k_BT \frac{\cos(2\pi\epsilon_F/\hbar\omega_c)}{\sinh(2\pi^2 M^* k_BT/\hbar eB)}. \tag{12.27}$$

Here we used

$$M^*\mu(T=0) = m^*\epsilon_F = \frac{1}{2}p_F^2, \tag{12.28}$$

which follows from the fact that the Fermi momentum p_F is the same for both dressed and undressed electrons.

In summary, (1) the SdH oscillation period is $\epsilon_F/\hbar\omega_c$, which is the same for the dHvA oscillations. This arises from the bunching of the quantum states. (2) The envelope of the oscillations exponentially decreases like $[\sinh(2\pi^2 M^* k_BT/\hbar eB)]^{-1}$. Thus, if the "decay rate" δ defined through

$$\sinh\left(\frac{\delta}{B}\right) \equiv \sinh\left(\frac{2\pi^2 M^* k_BT}{\hbar eB}\right), \tag{12.29}$$

is measured carefully, the magnetotransport mass M^* can be obtained *directly* through $M^* = e\hbar\delta/(2\pi^2 k_BT)$. This finding is quite remarkable. For example, the relaxation rate γ can now be obtained through Equation (12.2) with the measured magnetoconductivity. All electrons, not just those excited electrons near the Fermi surface, are subject to the E-field. Hence, the carrier density n appearing in Equation (12.2) is the total density of the dressed electrons. This n also appears in the Hall resistivity expression:

$$\rho_H \equiv \frac{E_H}{j} = \frac{ev_dB}{env_d} = \frac{B}{en}, \tag{12.30}$$

where the Hall effect condition $E_H = v_d B$, v_d = drift velocity, was used.

In the present theory, the two masses m^* and M^* are introduced corresponding to two physical processes: the cyclotron motion of the electron and the magnetotransport motion of the dressed electron. The dressed electrons are present whether the system is probed in equilibrium or in nonequilibrium as long as the system is subjected to a magnetic field. The presence of the c-particles can be checked by measuring the susceptibility or the heat capacity of the system. All (dressed) electrons are subject to the magnetic field, and hence, the magnetic susceptibility χ is proportional to the carrier density n, although the χ depends critically on the Fermi surface. This explains why the magnetic oscillations in the conductivity and the suceptibility are similar.

12.4 Heterojunction GaAs/AlGaAs

The heterojunction GaAs/AlGaAs is shown in Figure 12.4. The electrons are trapped in GaAs near the interface within a distance of the order of 100 Å. The Hall effect measurements are normally carried out in the geometry shown in Figure 12.5. The Hall field E_H is the Hall voltage V_H divided by the sample width W. The *Hall resistivity* ρ_H is defined by the ratio $\rho_H = E_H/j$, $E_H \equiv V_H/W$, where $j \equiv I/L = E/\rho$ (E = external field $\equiv V/L$) is the current density measured as the current I divided by the sample length L.

In 2001, Zudov et al. [5] observed two kinds of SdH-like oscillations in heterojunction GaAs/AlGaAs subject to microwave radiation. One kind periodic in B^{-1} which is visible with no microwave and which appears on the high field side ($B \sim 0.4$ T) and the second kind also periodic in B^{-1} appearing on the weak field side ($B \sim 0.2$ T) with a different period that exists only with the radiation. Mani et al. [6] and Zudov et al. [7] later found that the second kind of oscillation contains *zero resistance states* just like the quantum Hall (QH) states [8] occurring at higher fields ($B \sim 5$ T) but *without* the *Hall resistance plateaus*. Figure 12.6 represents the data after Mani [9] for the resistance R_{xx} versus the reduced inverse magnetic field ${B'}^{-1}$. Note that the curves with (w/) and without (w/o) microwaves intersect at 3, 4, 5, ..., and the two kinds of oscillations visible are equally spaced in ${B'}^{-1}$.

A set of the prominent oscillations occur for the sample with no radiation. We concentrate on this case. Figure 12.6 show that the resistance $R_{xx} \equiv E/I$ linearly decreases with B^{-1} in very low fields. This behavior can be explained by using Equation (12.2) as follows. For high-purity samples at very low

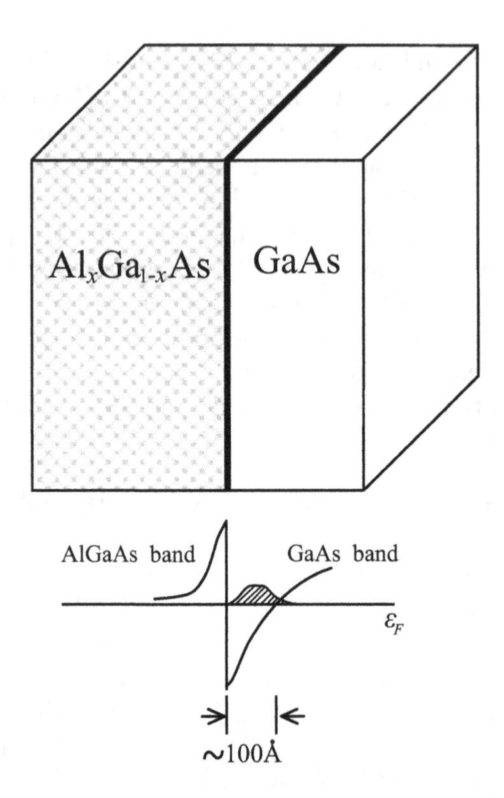

Figure 12.4: The electrons are localized in less than 100 Å near the interface GaAs/AlGaAs. The shaded area represents the electron distribution.

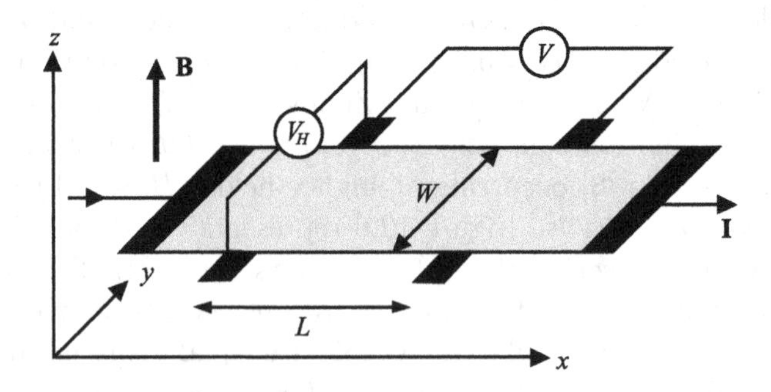

Figure 12.5: The Hall effect measurements. The Hall voltage V_H is generated perpendicular to the current **I** and the magnetic field **B**.

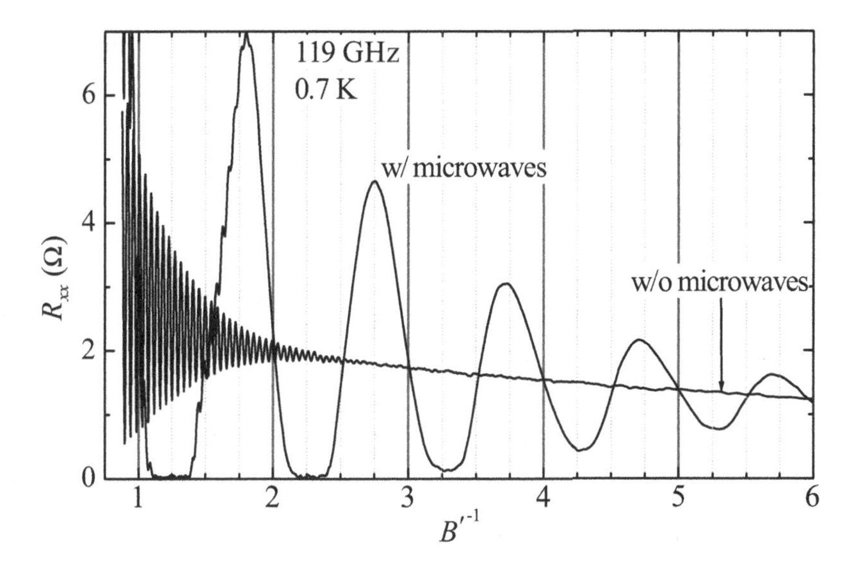

Figure 12.6: A plot of resistance R versus the reduced inverse magnetic field B' after Mani [9]. See Ref. 9 for the actual reduction.

temperatures ($\sim$ 0.7 K) the impurity and phonon scatterings are negligible. By the *energy-time uncertainty principle* the dressed electron can spend a short time at an upper LL and come back to the ground LL with a different guiding center, thus causing a guiding center jump. We assume that the relaxation rate γ is the *natural line width* arising from the LL separation divided by $\hbar$, that is, the cyclotron frequency ω_c:

$$\gamma = \omega_c \equiv \frac{eB}{m^*}. \tag{12.31}$$

This generates the desired B^{-1} dependence.

We fitted Mani's data in Figure 12.6 with

$$\begin{aligned} R &= A + Bx + \frac{E\cos(2\pi Cx) + F}{\sinh(Dx)}, \\ A &= 2.3, \quad B = -0.18, \quad C = 23, \quad D = 3.1, \\ E &= 22.0, \quad F = 7.0. \end{aligned} \tag{12.32}$$

The fits agree with the data within the experimental errors. Using

$$Dx = \frac{\delta}{B}, \tag{12.33}$$

we obtain

$$M^* = 0.30m. \tag{12.34}$$

This value is about 4.5 times greater than the cyclotron mass m^*,

$$m^* = 0.067m, \tag{12.35}$$

which is known from other experimental data. This difference arises as follows.

In the *cyclotron motion* the electron with the effective mass m^* circulates around the magnetic fluxes. Hence, the cyclotron frequency ω_c is given by eB/m^*. The guiding center (dressed electron) moves with the magnetotransport mass M^*, and therefore, this M^* appears in the hyperbolic sine term in Equation (12.27).

In 1952 Dingle [10] developed a theory of the dHvA oscillations. He proposed to explain the envelope behavior in terms of a *Dingle temperature* T_D such that the exponential decay factor be

$$\exp\left[\frac{-\lambda(T+T_D)}{B}\right], \qquad \lambda = \text{constant}. \tag{12.36}$$

Instead of the modification in the temperature, we introduced the magnetotransport mass M^* to explain the envelope behavior. The susceptibility χ is an equilibrium property, and hence, χ should be calculated without considering the relaxation mechanism. In our theory, the envelope of the oscillations is obtained by taking the average of the sinusoidal density of states with the Fermi distribution of the dressed electrons. There is no place where the impurities come into play. The theory may be checked by varying the impurity density. Our theory predicts little change in the clearly defined envelope. The scattering will change the relaxation rate γ and the magnetoconductivity for the center of the oscillations through Equation (12.2).

The field-dressed electron is the same entity as the *composite particle* (c-particle) having an electron and fluxons in the theory of the *Quantum Hall Effect* (QHE) [8]. We shall briefly describe this effect here. Figure 12.7

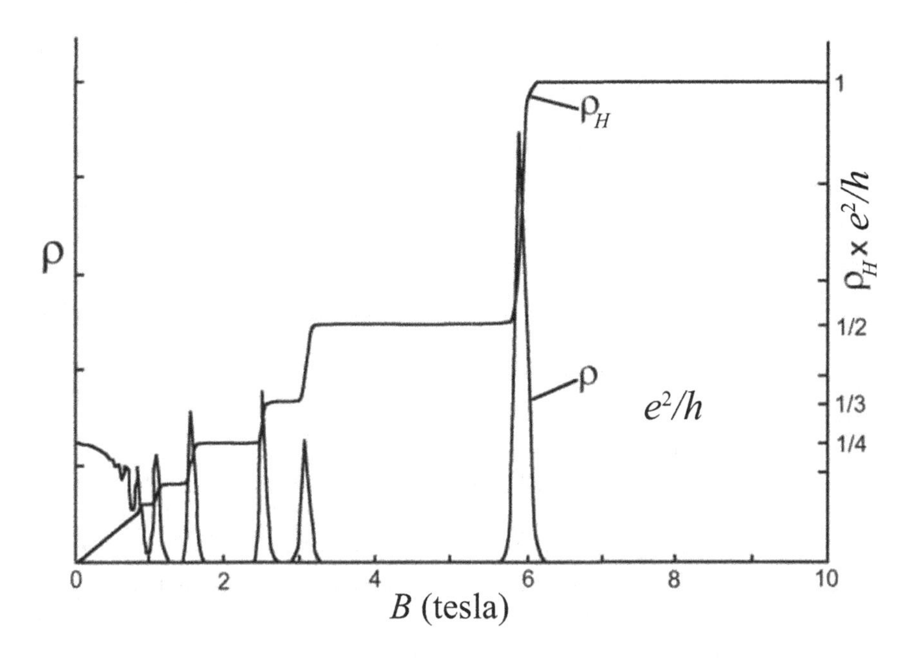

Figure 12.7: Observed QHE in heterojunction GaAs/AlGaAs at 60 mK, after Tsui [11]. The Hall resistivity ρ_H and the resistance ρ are shown as a function of the magnetic field B in tesla.

represents the data reported by Tsui [11] for the Hall resistivity ρ_H and the resistivity ρ in GaAs/AlGaAs at 60 mK. The quantum Hall (QH) states at the LL *occupation ratio*, also called the *filling factor*, $\nu = 1, 2, \ldots, 5$ are visible in this figure. Clearly, each QH state with the Hall resistivity plateau (horizontal stretch) is *superconducting* (zero resistance). The plateau heights are *quantized* in units of h/e^2, h = Planck constant, e = electron charge (magnitude). Each plateau is material- and shape-independent, indicating the fundamental quantum nature, the boundary-condition independence, and the stability of the QH state. The quantum statistics of a compsite particle with respect to the center-of-mass motion is given by the *Ehrenfest–Oppenheimer–Bethe's rule* [12] that the c-particle moves as a fermion (boson) if it contains an odd (even) number of the *elementary fermions*. Applying this rule to the dressed electron composed of an electron and a fluxon, we obtain a c-boson if we regard the fluxon as an elementary fermion [13]. The countability of the fluxons is known as the *flux quantization* [see Equation (11.11)]. Bosons, such as photons and phonons, can be absorbed by materials, while fermions, such as electrons and protons, cannot. Since the magnetic fluxes cannot disappear at a sink, the *fluxon*, the *quantum of the magnetic field*, is by postulate a fermion with zero mass and zero charge.

Mani [9] observed the (diagonal) resistance R_{xx} and the Hall resistance R_{xy} defined by

$$R_{xy} \equiv \frac{E_H}{J}, \qquad J = \text{current}, \tag{12.37}$$

in heterojunction GaAs/AlGaAs with and without radiation (50 GHz) up to 6 T (tesla) at 0.5 K. His data are reproduced in Figure 12.8. The abscissas are the magnetic field B (T) in (a) and B (mT) in (b). In Figure 12.8 (a) we see that the resistance R_{xx} vanishes (*superconducting*) at 4.2, 3.2, ... (B in T), accompanied by the horizontal stretches in the Hall resistivity R_{xy}. This is the main signature of the QHE. The numbers 3 and 4 in Figure 12.8 (a) are the *Landau level occupation numbers* ν, which are explained below.

According to Onsager's flux quantization represented by Equation (11.11), the magnetic fluxes can be counted in units of $\Phi_0 = e/h$. We first consider the special case in which the number of fluxons is equal to the number of the electrons in the sample. Each dressed electron (composite particle) then carries one fluxon. This condition in the theory of the QHE is represented by the LL occupation number $\nu = 1$. Note this condition is realized at different fields for different samples. At $\nu = 1$, all of the electrons occupy the lowest LL. Consider now the case in which the field is reduced by half compared with the case at $\nu = 1$. The LL degeneracy is reduced. The electrons, therefore, fill the lowest two (2) LLs. This condition is represented by $\nu = 2$. At $\nu = N$ (integer), the lowest N LLs are filled by electrons. If the magnetic field is high such that the number of fluxons equals twice the number of the electrons, then this condition is represented by $\nu = 1/2$, where each dressed electron carries two fluxons. The LL occupation number ν is also called the *filling factor* (of the lowest LL).

Let us now examine the data in Figure 12.8. The main feature in Figure 12.8 (a) is the integer QHE at $\nu = 3, 4, ...$, which are caused by the *Bose-Einstein condensation* of the c-bosons, each with one fluxon. The deep dips at $\nu = 5$ and higher are separated by broad maxima at $\nu = 11/2, 13/2, ...$, which are due to the c-fermions, each with two fluxons. The two smaller dips

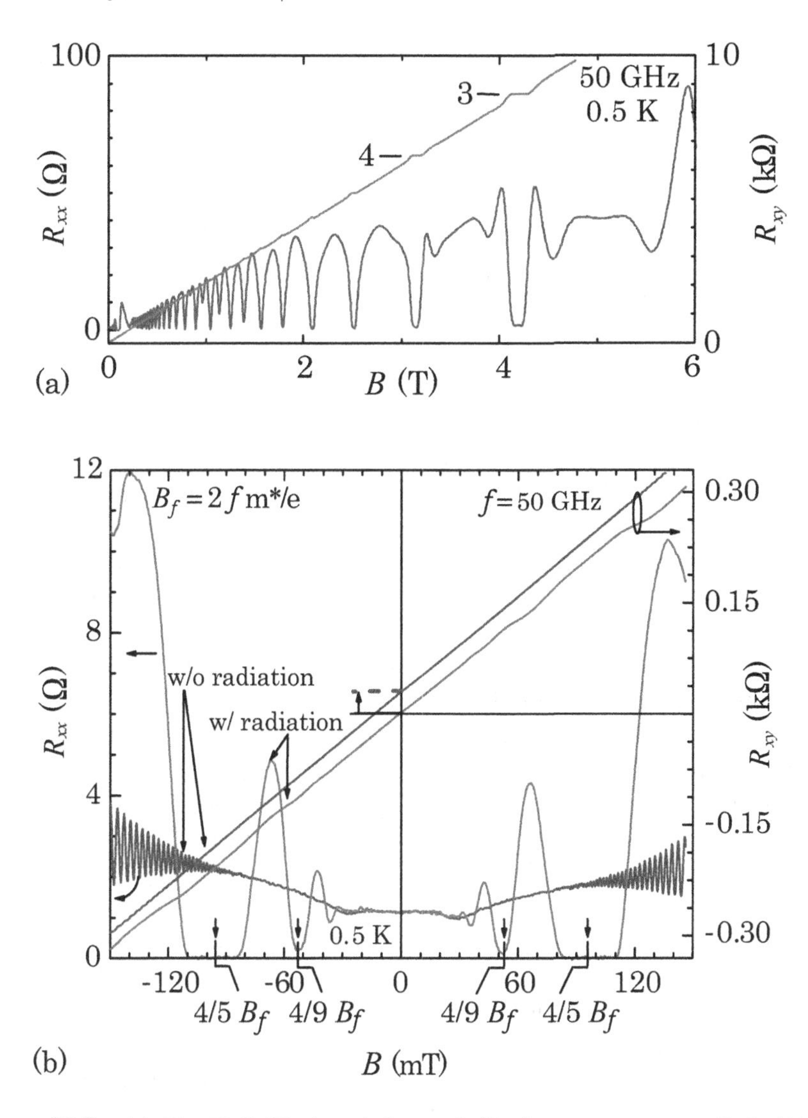

Figure 12.8: (a) The Hall (R_{xy}) and diagonal (R_{xx}) resistances in a GaAs/AlGaAs heterostructure under excitation at 50 GHz, after Mani [9]. Quantum Hall effects (QHE) at $\nu = 3, 4$ are visible for high B [$\sim$ T(tesla)], where R_{xx} vanishes accompanied by plateaus in R_{xy}. For low fields (0.2 T or below), new types of QHE appear with radiation. (b) Data over low magnetic fields obtained both with (w/) and without (w/o) excitation at 50 GHz. Here, radiation-induced vanishing resistance about $(4/5)B_f$ does not induce plateaus in the Hall resistance, unlike in the usual QHE.

appearing between $\nu = 4$ and $\nu = 3$ arise from the bosonic states at $\nu = 11/3$ and $10/3$. The *new* states for field $B < 0.25$ T are the radiation-induced *superconducting states* discovered by Mani et al [6]. The expanded view of the low-field data are shown in Figure 12.8 (b), where two zero-resistance states marked by arrows are clearly visible for the curve with radiation. We also notice the distinctive SdH oscillations in the curve without radiation at $B = 80$ mT and higher.

Jain introduced the *effective magnetic field* [14]

$$B^* \equiv B - B_\nu \equiv B - \left(\frac{1}{\nu}\right) n_e \left(\frac{h}{e}\right) \tag{12.38}$$

relative to the standard field B_ν for the c-fermion at the even-denominator fraction ν. We extend this idea to the bosonic (odd-denominator) fraction. This means that the CM of c-particle (guiding center) moves fieldfree at the exact fraction, where all fluxons in the system are attached to the c-particles. The excess (or deficit) of the magnetic field is simply the effective magnetic field B^*.

The QH state at $\nu = N$ is similar to that at $\nu = 1$. All states at $\nu = N$ are generated by the BEC of the c-bosons, each with one fluxon. Only the c-boson density is smaller with higher N, and hence, the plateau length is shorter as seen in Figure 12.8 (a).

The QH states at $\nu = N$ and $N + 1$ have different bosonic configurations and cannot be smoothly connected. In between there is the fermionic state at $\nu = N + 1/2$, which exhibits a resistance. The resistance maxima at $\nu = 4$ and greater N in Figure 12.8 (a) are caused by the c-fermions, each with two fluxons while the minima are due to the (partially) condensed c-bosons. By lowering fields B (< 0.25 T) these maxima and minima smoothly become the SdH oscillations [see Figures 12.8 (b) and 12.4]. In the low field limit the c-bosons disappear. The binding energy *and* the entropy are greater for the c-fermions than for the c-boson. Thus, from the Helmholtz free-energy consideration the c-fermions dominate in the lowest fields.

At the exact fraction $\nu = (N + 1)/2$, all fluxons are attached to the electrons, and the excess (or deficit) of the magnetic field appears as the (Jain's original) effective field B^* defined in Equation (12.38). This effective field B^* should enter as the magnetic force term in Equation (12.23):

$$e(\mathbf{v} \times \mathbf{B}^*) \cdot \frac{\partial \phi}{\partial \mathbf{\Pi}}. \tag{12.39}$$

At the low fields, the magnitude of the effective field B^* is less than half the distance (field) of the neighboring minima:

$$|B^*| \leq \frac{1}{2}(B_N - B_{N+1}). \tag{12.40}$$

Hence, the magnetic force term is negligible compared with the electric force term $e\mathbf{E} \cdot \partial\phi/\partial\mathbf{\Pi}$.

The effective mass M^* estimated from the QH state at $\nu = 1/2$ has the values $M^* \sim 0.82 - 1.0m$ [15]. Thus, our obtained value 0.30 is closer to these values. If we assume that the c-particle is generated by the phonon exchange, positive and negative c-particles are created simultaneously. We may assign $0.30m$ to the majority carrier mass. The minority carrier has a heavier mass. Thus, the observed value $(0.8 - 1.0m)$ may arise from these two carriers of different masses.

More detailed discussions of the QHE are given in book 2 [16]. The oscillations in R which appear with microwaves are bosonic and will be discussed separately.

Chapter 13

Cyclotron Resonance

Cyclotron resonance will be treated in this chapter. The most direct probe of the Fermi surface can be made by observing the cyclotron resonance. A magnetic field is applied to a pure sample at liquid helium temperatures. The sign of the charge carrier can be determined by using the circularly polarized lasers. The data are analyzed in terms of Shockley's formula or its simplified version. Most often the effective masses for a conductor are determined directly after simple analyses.

13.1 Introduction

"Electrons" ("holes") are fermions having negative (positive) charge $q = -(+)e$ and positive effective masses m^* that respond to the Lorentz force $\mathbf{F} = q(\mathbf{E} + \mathbf{v} \times \mathbf{B})$. These conduction electrons ("electrons," "holes") are generated at finite temperatures in a metal, depending on the curvature of the metal's Fermi surface.

The operating principle of *cyclotron resonance* (CR) is simple. Take a quasifree electron model. A *quasifree electron* by definition has an effective mass m^* but otherwise behaves similarly to a free electron. If a magnetic field $\mathbf{B}$ is applied, all electrons with the dispersion relation

$$\epsilon = \frac{1}{2m^*}(p_1^2 + p_2^2 + p_3^2) \tag{13.1}$$

will move, each electron keeping the magnetic flux inside intact. These electrons have in general different *circulating speeds*:

$$v_\perp \equiv (v_1^2 + v_2^2)^{1/2}. \tag{13.2}$$

But all of the electrons under the magnetic field move with the *same* angular frequency

$$\omega_c \equiv \frac{qB}{m^*}, \tag{13.3}$$

called the *cyclotron frequency*. If a *monochromatic radiation* (microwave) with the matching frequency ω

$$\omega = \omega_c, \tag{13.4}$$

is applied, a significant absorption peak can be observed. The peak position (frequency) for the quasifree electron model directly gives the value of the effective mass $m^* = qB/\omega_c$ according to Equation (13.3).

To actually observe the resonance peak, certain experimental conditions must be met: (1) the sample must be pure, (2) the experiments should be performed at liquid helium temperatures, and (3) the operating frequency ω must be reasonably high. These conditions ensure that the resonance width (frequency) arising from the electron-impurity and the electron-phonon scatterings is much smaller than the peak frequency.

13.2 Cyclotron Resonance in Ge and Si

In 1955, Dresselhaus, Kip, and Kittel (DKK) [1] reported a full account of the *first* observation of the cyclotron resonance in germanium (Ge) and silicon (Si). They found "electrons" and "holes" in both semiconductors and that the resonance peaks change with the orientation of the field relative to the crystal axes. After analyzing the data in terms of the DKK formula, Equation (13.7), they obtained the "electron" ellipsoidal Fermi surfaces with the effective masses

$$(m_1, m_2, m_3) = (0.082,\ 0.082,\ 1.58)m \quad \text{for Ge}, \tag{13.5}$$

$$(m_1, m_2, m_3) = (0.19,\ 0.19,\ 0.97)m \quad \text{for Si}. \tag{13.6}$$

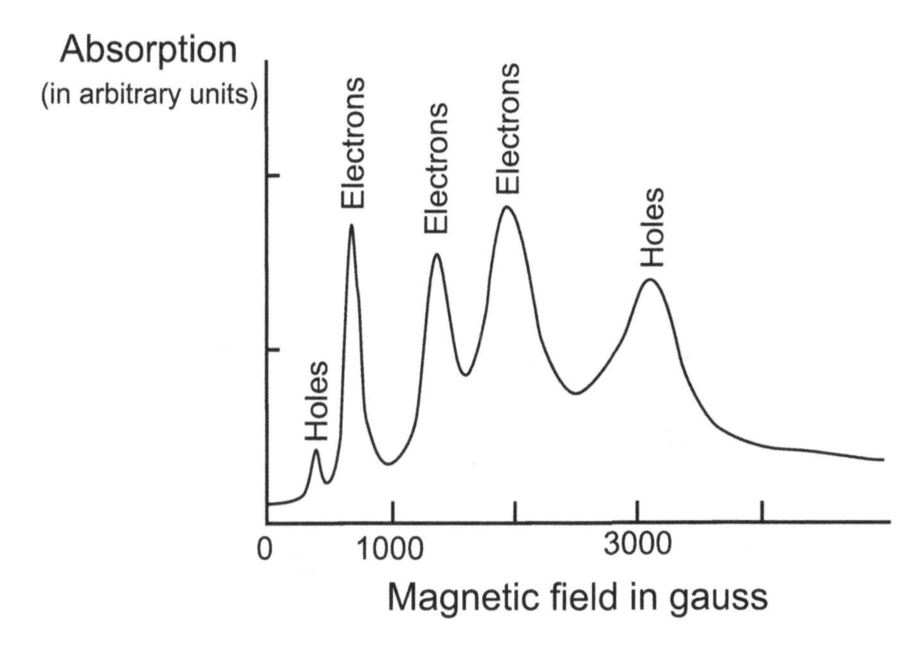

Figure 13.1: Cyclotron resonance absorption for Ge at $\nu = 24 \times 10^9$ cycles sec^{-1} at $T = 4$ K, after DKK [1].

This was a historic event in solid-state physics. The effective masses $m_i^* \equiv m_i$, i =1, 2, 3, are material constants distinct from the gravitational electron mass m. The effective mass enters as a phenomenological parameter in the quasifree electron model of a metal. The effective mass appears in the kinetic-theoretical discussion of the electron transport (see Section 4.1). It also appears in the discussion of the low-temperature heat capacity behavior. The fact that the effective mass m^* can be measured directly by the CR experiments is most remarkable. The Fermi surface generally is nonspherical in metals (semiconductors), and hence, the electron transport is anisotropic.

Figure 13.1 represents typical experimental data for Ge obtained by DKK. In the abscissa, the magnetic field rather than the microwave frequency is taken. In the experimental runs, the magnetic field is varied while the frequency is held constant. The charge carriers can be identified by applying circularly polarized microwaves (see Figure 13.1). If the carrier is an "electron" ("hole") and circulates around the magnetic field counterclockwise

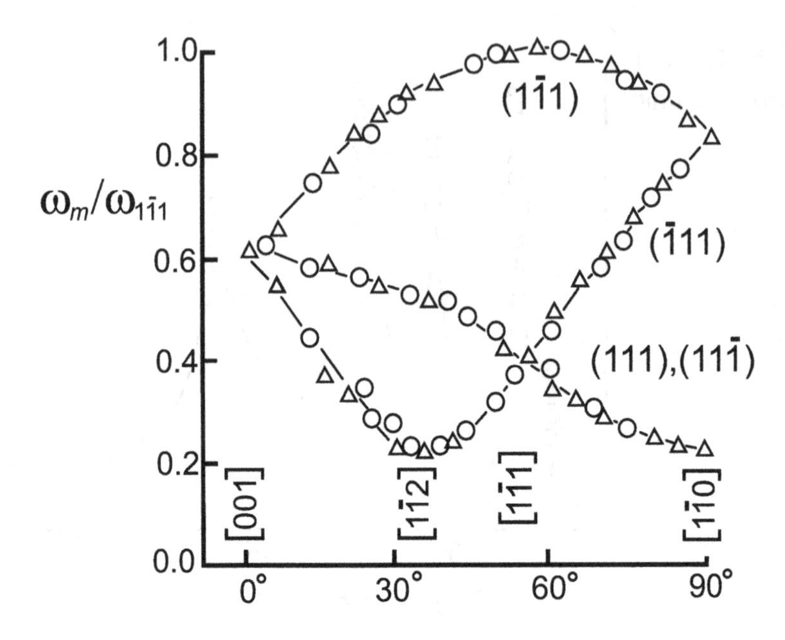

Figure 13.2: The orientation dependence of the resonance maxima ω_m in a (110) plane for "electrons" in Ge. Theoretical curves are obtained from Equation (13.7) with $m_t = 0.082m$ and $m_l = 1.58m$, and experimental data are reproduced after DKK [1].

(clockwise), it can absorb energy from the microwave whose polarization vector circulates counterclockwise (clockwise). If the field orientation is varied, all peak positions were found to move. In DKK's controlled experiments, the magnetic field is rotated in a (110) plane from [001] to [1$\bar{1}$0]. The reduced resonance frequency versus the field angle is shown in Figure 13.2. The solid lines are theoretical curves based on the *DKK formula*:

$$\omega = \left(\omega_t^2 \cos^2\theta + \omega_t\omega_l \sin^2\theta\right)^{1/2}, \quad \omega_t \equiv \frac{eB}{m_t}, \quad \omega_l \equiv \frac{eB}{m_l}. \tag{13.7}$$

Notice the excellent agreement between the theory and the experimental data.

Formula (13.7) can be derived with the assumption of a *spheroidal dispersion relation* represented by

$$\epsilon = \frac{p_1^2}{2m_t} + \frac{p_2^2}{2m_t} + \frac{p_3^2}{2m_l}. \tag{13.8}$$

First, let us take wurtzite (ZnS), which has highly anisotropic transport properties [2], and which may be thought to have a set of conducting planes perpendicular to the c-axis, each plane containing a hexagonal array of ions (Zn^{2+} or S^{2-}). If a magnetic field $\mathbf{B}$ is applied along the c-axis, the conduction electrons may orbit around the field $\mathbf{B}$ and passing through a series of + ions forming hexagons of various sizes. The smallest hexagonal orbital is similar to the molecular orbital for C_6H_6 originally proposed by Hückel for this molecule [3]. The quantum energies for the closed hexagonal orbitals are expected to be of the form:

$$\left(n + \frac{1}{2}\right)\hbar\omega_c\,, \qquad \omega_c = \frac{eB}{m_t}, \tag{13.9}$$

where ω_c is the cyclotron frequency. These molecular orbitals are highly degenerate since the centers of the hexagons can be anywhere in each plane. Because the crystal-atom array and, therefore, the crystal potential is periodic in the z-direction (c-axis), the electron is unlocalized in this direction, and the energy E should depend on the wave number k_z by Bloch's theorem. We may then represent the energy E by

$$E = E_0 + \frac{\hbar^2 k_z^2}{2m_l} + \left(n + \frac{1}{2}\right)\hbar\omega_c, \qquad E_0 = \text{constant}. \tag{13.10}$$

The effective masses m_t and m_l appearing in Equations (13.9) and (13.10) are different from each other. The transverse mass m_t is equal to the effective mass assocated with the hexagonal-base plane in the field-free case. In contrast, the longitudinal mass m_l is associated with the motion along the c-axis. The energy levels represented by Equation (13.10) can be regarded as those for a fictitious particle with charge e and anisotropic masses (m_t, m_l), characterized by the Hamiltonian

$$H = \frac{1}{2m_t}[(p_1 - qA_1)^2 + (p_2 - qA_2)^2] + \frac{1}{2m_l}(p_3 - qA_3)^2, \tag{13.11}$$

where $\mathbf{p} \equiv \hbar k$ is the momentum and $\mathbf{A}$ is the vector potential, which generates a constant magnetic field $\mathbf{B}$ such that $\mathbf{B} = \nabla \times \mathbf{A}$. We now look for a cyclotron frequency ω_c by considering the classical motion of this fictitious particle.

According to standard Hamiltonian mechanics, the velocity components

(v_1, v_2, v_3) are given by

$$\begin{aligned} v_i &= \frac{dx_i}{dt} = \frac{\partial H}{\partial p_i} = \frac{1}{m_t}(p_i - qA_i), \qquad i = 1, 2, \\ v_3 &= \frac{1}{m_l}(p_3 - qA_3). \end{aligned} \tag{13.12}$$

These components move following Newton's equations of motion:

$$\begin{aligned} m_t \frac{dv_i}{dt} &= q(\mathbf{v} \times \mathbf{B})_i, \qquad i = 1, 2, \\ m_l \frac{dv_3}{dt} &= q(\mathbf{v} \times \mathbf{B})_3, \end{aligned} \tag{13.13}$$

which are linear homogeneous differential equations. We introduce $v_j(t) = e^{-i\omega t} v_j$ in Equations (13.13) and assume that not all amplitudes v_j are zero. We then obtain a (determinant) *secular equation* (Problem 13.2.1)

$$\begin{vmatrix} i\omega m_t & -qB\cos\theta & qB\sin\theta\sin\phi \\ +qB\cos\theta & i\omega m_t & -qB\sin\theta\cos\phi \\ -qB\sin\theta\cos\phi & qB\sin\theta\sin\phi & -i\omega m_l \end{vmatrix} = 0, \tag{13.14}$$

the solution of which yields Equation (13.7).

Notice that the frequency ω depends neither on the speed v nor on the energy of the particle. If the CR experiments are performed, then the resonance maximum position ω_m is given by $\omega_m = \omega$, and the maximum should be sharp in the absence of scatterers since all the electrons resonate at the same frequency. No experiments for this crystal (wurtzite) appears to be available at the present time.

The idea of hexagonal cyclotron orbitals can be extended to the diamondlike crystal Ge. This crystal has hexagonal-base planes in directions perpendicular to $[111]$, $[1\bar{1}1]$, $[11\bar{1}]$, and $[\bar{1}11]$ or $\langle 111 \rangle$. The electrons may move easily on the hexagonal planes. If a high magnetic field is applied in a general direction, then there should be four CR peaks. As the orientation relative to the crystal is varied, the peak positions change, each following Equation (13.7). Dresselhaus, Kip, and Kittel [1] measured the CR maxima for Ge at 4 K for magnetic field directions in a (100) plane, plotted the data with the apparent effective mass $m^* = eB/\omega_m$ as the ordinate, and analyzed them based on the assumed set of spheroidal energy surfaces represented by Equation (13.8). They obtained an excellent fit with $m_t = 0.082m$ and $m_l = 1.58m$ (see Figure 13.2). In the actual experimental range of directions, there are three resonance maxima because the maxima associated with two hexagonal-base planes, (111) and $(11\bar{1})$, coincide. These maxima are more robust than the other two maxima. See Figure 13.1, where the "electron" peak at the highest field corresponds to the doubly degenerate peak.

The theoretical formula for the resonance frequency ω used in their analysis is identical to Equation (13.7). We derived this formula a little differently by introducing the *cyclotronic plane* (CP) in which the electron circulates. The excellent agreement between theory and experiment means that Ge's "electron" Fermi surface is a spheroid with the major axes pointing along $\langle 111 \rangle$, having the effective masses $(m_t,\ m_l)$ given in Equation (13.5). The cyclotron radius l corresponding to the lowest Landau states with $n = 0$ is given by $l = (\hbar/eB)^{1/2}$, which is about 180 nm at $B = 0.02$ T (typical experimental value). If the set of four spheroidal surfaces is located at the centers of the Brillouin-boundary hexagons whose normal points in the $\langle 111 \rangle$ directions, then the three lines of numerical fits are obtained *all at once* with the choice of the mass parameters $(m_t,\ m_l)$.

Ge and Si share the same *diamond-lattice structure*. However, the CR peaks observed in Si were found to be very different from those in Ge. The data obtained by DKK [1] are shown in Figure 13.3, where the apparent effective mass ratio $m^*/m \equiv \omega/\omega_m$ is plotted as a function of the field angle from [001] in a (110) plane. This difference arises from the fact that the CPs are different, as explained below.

First, consider a *zinc blende* (ZnS) in which Zn's and S's occupy the diamond structure such that each Zn (S) is surounded by four S (Zn); the sublattice of Zn (or S) form a FCC lattice. If we disregard the species, the zinc blende (lattice) structure becomes the diamond structure.

We now consider Ge forming a diamond structure. As we saw earlier the CPs in Ge are the hexagonal-base planes $\{111\} = (111)$, $(1\bar{1}1)$, $(11\bar{1})$ and $(\bar{1}11)$, which are perpendicular to the directions $[111]$, $[1\bar{1}1]$, $[11\bar{1}]$, and $[\bar{1}11]$, respectively. We now show that the CPs in Si are planes $\{100\} = (100)$, (010), and (001). The plane (100) containing the face-centered square can be viewed as the square lattice with the sides directed in $[1\bar{1}0]$ and $[\bar{1}10]$. Hence the electron motion in the plane is isotropic and can be characterized by a

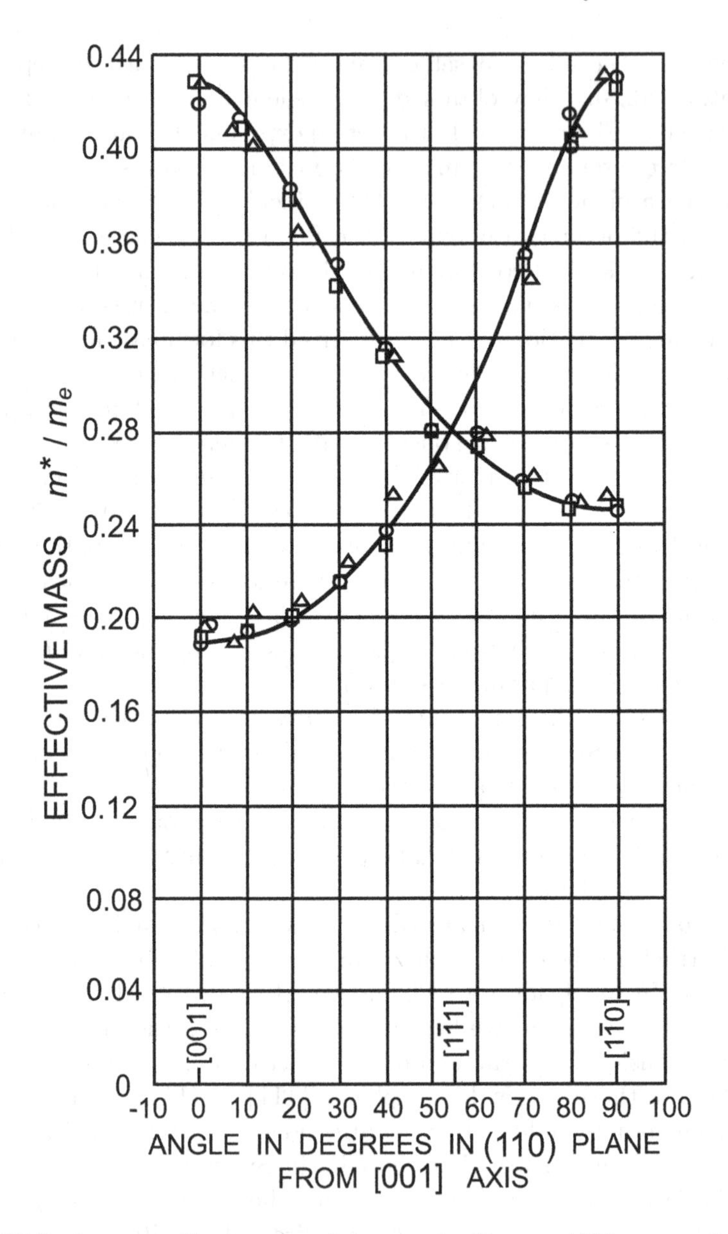

Figure 13.3: Apparent effective mass of electrons in silicon at 4 K for magnetic field directions in a (110) plane after DKK [1]; the theoretical curves are calculated from Equation (13.15), with $m_l = 0.98m$; $m_t = 0.19m$.

single mass m_t. The motion perpendicular to the plane (110) is characterized by a different mass m_l, which should be heavier than m_t since the periodicity length in this direction is greater than that in the plane (001), and hence the electrons move less easily. Thus, the electron energy E for Si can be represented by the same equation (13.10) as for Ge. The number of peaks and the orientation dependence of each peak can be fitted, based on Equation (13.7), if we assume that *three* spheroidal Fermi surfaces are located at the centers of the Brillouin boundary whose normal vectors point in $\langle 100 \rangle$ directions. For the experimental geometry only two curves appear in Figure 13.3. The absorption peak coming from the two CPs, (100) and (010), is degenerate and is more robust than that arising from the other plane (001).

Dresselhaus, Kip, and Kittel reported [1] that (1) there are two or more CR peaks for "holes" in Si, and (2) the peak at the greatest frequency (the smallest effective mass) is independent of the field orientation, while the other peaks depend on the orientation (see Figure 13.4). Similar behaviors are also observed for "holes" in Ge.

The light "holes" (the greatest CR frequency) for Si and Ge have the effective masses [4]

$$m_1 = m_2 = m_3 = 0.16\, m \quad \text{for light "holes" in Si,} \tag{13.15}$$

$$m_1 = m_2 = m_3 = 0.042\, m \quad \text{for light "holes" in Ge,} \tag{13.16}$$

which describe the orientation-independent peaks, when the field direction is tilted in the (110) plane from [001] to $[1\bar{1}0]$ direction. See Figure 13.4. We may interpret this behavior with the assumption that the CPs are $\{100\}$ planes, which contain three equivalent planes (100), (010), and (001) just like the SC lattice. Each plane however contains more ions in the FCC lattice than in the SC lattice. This feature will introduce a new set of modes as explained below.

Dresselhaus, Kip, and Kittel fitted the orientation-dependent CR peaks in terms of the fluted closed-energy orbit shown in Figure 13.5. But the fits near 50° are not very good, as we see in Figure 13.4. The fluted closed orbit in the k-space contains sections with positive curvatures *and* sections with negative curvatures, and hence, it is not the physically executable orbit (no cyclotron frequency) as explained earlier in Section 10.2.

The "hole" and the "electron" are similar except for the different charge

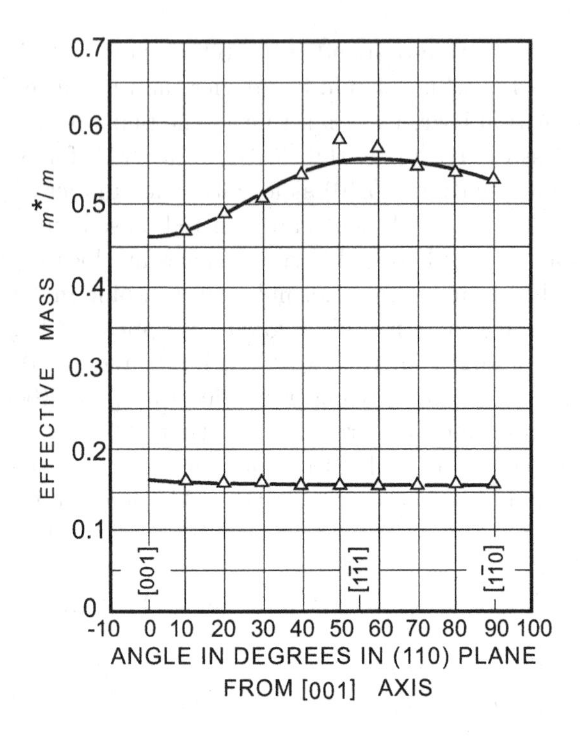

Figure 13.4: Effective mass of "holes" in silicon at 4° K for magnetic field directions in a (110) plane, after DKK [1].

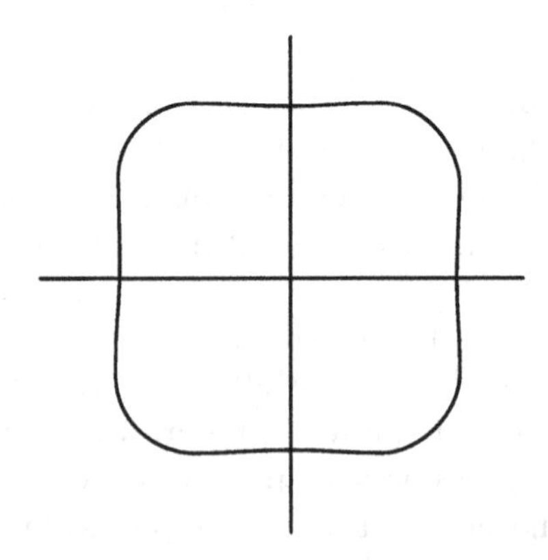

Figure 13.5: A fluted constant energy orbit in the (100) plane of k-space.

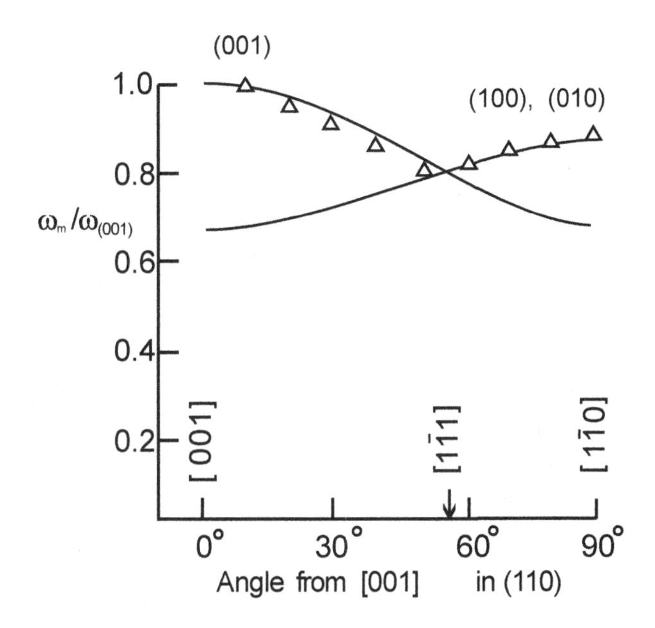

Figure 13.6: The orientation dependence of reduced resonance maxima ω_m for "holes" in Si, reproduced after DKK [1]. The magnetic field is rotated in a (110) plane and the field angle measured from the [001] orientation is taken as the abscissa. The theoretical curves (solid lines) are obtained from Equation (13.7) with $m_t = 0.46m_e$ and $m_l = 1.03m_e$.

signs. Therefore, they should be treated in a symmetric manner. We shall present a new theory here. In Figure 13.6 we replot the angle-dependent CR data with the peaks ratio ω_m/ω_{001} as the ordinate. The field is rotated in a (110) plane, and the field angle is measured from the [001] orientation. The data are indicated by triangles Δ, and the theoretical curve is fitted using DKK formula (13.7) with

$$m_t = 0.46\,m,\;\; m_l = 1.03\,m \;\; \text{for heavy "holes" in Si.} \tag{13.17}$$

Our fits in Figure 13.6 are distinctly better than DKK's fits shown in Figure 13.4, top curve. We have assumed that the CPs are $\{100\}$ planes. There are only two curves since the two CPs, (100) and (010), are degenerate in this tilt geometry. At the $[1\bar{1}1]$ orientation the two theoretical curves coincide since the angles between the field and any one of the planes are the same.

Both angle-dependent and -independent peaks for the "holes" are fitted with the assumption that the CPs are $\{100\}$ planes. This will now be explained. We mark the cube-edge atoms of the unit FCC lattice by open circles $\circ$ and the face-center atoms on the top and bottom faces by closed circles $\bullet$. See Figure 13.7 (a). Both atoms $\circ$ and $\bullet$ form separately SC lattices of side-length a as shown in Figure 13.7 (b). The face-center atoms on the sides, marked by semiopen circles $\odot$ in Figure 13.7 (a), form a tetragonal lattice of base length $a/\sqrt{2}$ and height a as shown in Figure 13.7 (c).

Let us now discuss the orientation-dependent CR peaks in Figure 13.6. The CPs are $\{100\}$, the same as those for the orientation-independent CR peaks. But the peaks are generated from the tetragonal sublattice shown in Figure 13.7 (c). The side length of the base square is $a/\sqrt{2}$, while the height length is a. The effective mass $m_t = 0.46m$ for the base-plane motion should be, and is, smaller than the longitudinal effective mass $m_l = 1.03m$.

In summary, light and heavy "holes" based on the SC and tetragonal sublattices, respectively, produce the orientation-independent and -dependent CR peaks.

For Ge the CR peaks for heavy "holes" are orientation-dependent just as heavy "holes" are in Si. After similar analyses, we obtain the following effective masses:

$$m_t = 0.29m, \; m_l = 0.78m \quad \text{for heavy "holes" in Ge.} \tag{13.18}$$

These masses are associated with the tetrahedral sublattice in Figure 13.7 (c) with the CP in $\{100\}$ planes.

We have decomposed the FCC lattice into two SC sublattices and a tetragonal sublattice. Such decomposition should be helpful in discussing the normal modes of the FCC lattice. The decomposition into sublattices is also useful in the discussion of the band structures of metals and semiconductors.

Let us cite a few examples. Consider a BCC lattice. This lattice can be decomposed into two SC sublattices, one made of the cube-edge atoms ($\circ$) in the unit cell and another made of the body-center atoms ($\bullet$) as shown in Figure 13.8. We may deduce that Na (BCC) should have a spherical Fermi surface expected of a SC (idealized) lattice.

Binary compounds such as InSb and GaAs have a zinc blende (ZnS) lattice structure. In this lattice Zn^{2+} form a FCC sublattice and S^{2-} also form a FCC sublattice with each S^{2-} surrounded by four Zn^{2+}. The FCC sublattice may further be decomposed into SC and tetragonal lattices. The energy bands of a zinc blende lattice compound is most often discussed by using the cubic lattice languages.

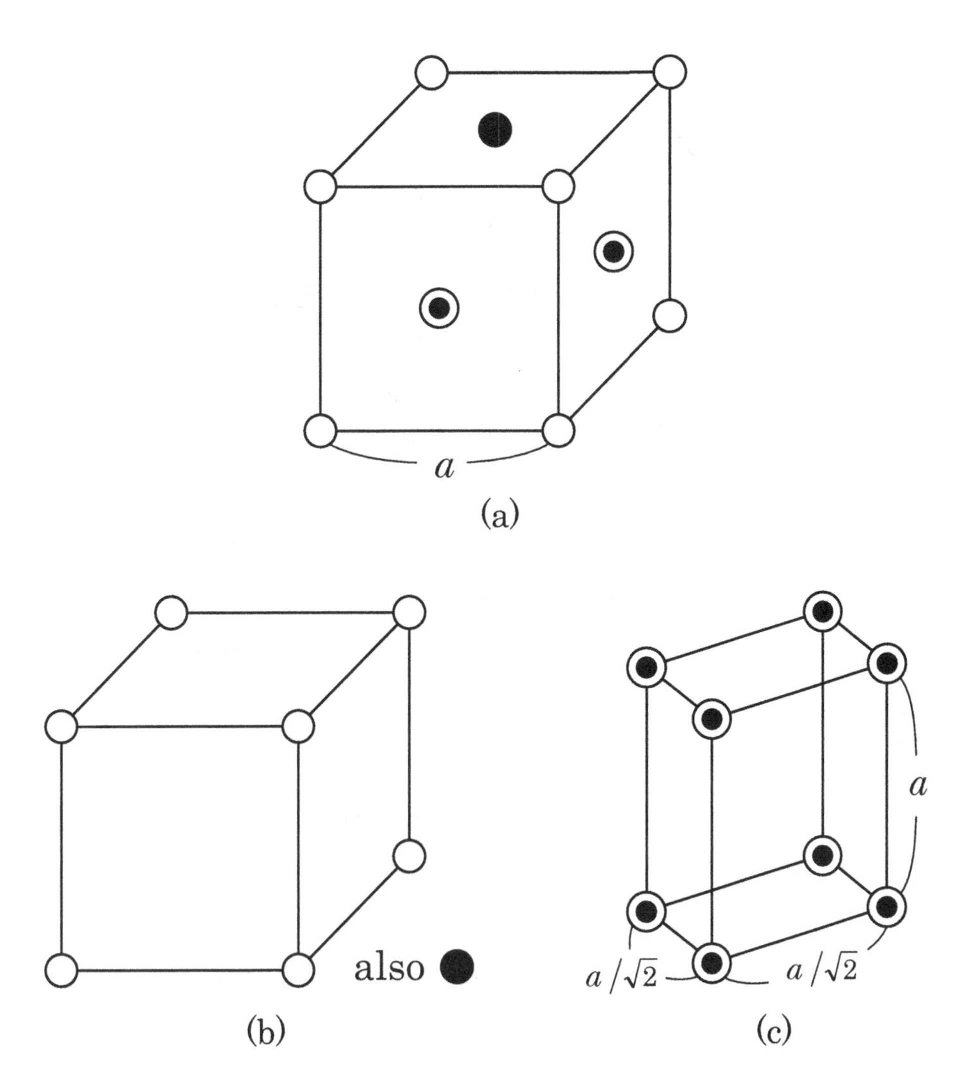

Figure 13.7: (a) The unit FCC lattice. (b) The cube-edge atoms (∘) form a SC sublattice. The face-center atoms (•) on the top and bottom faces also form a SC sublattice. (c) The face-center atoms (◉) on the sides form a tetragonal sublattice.

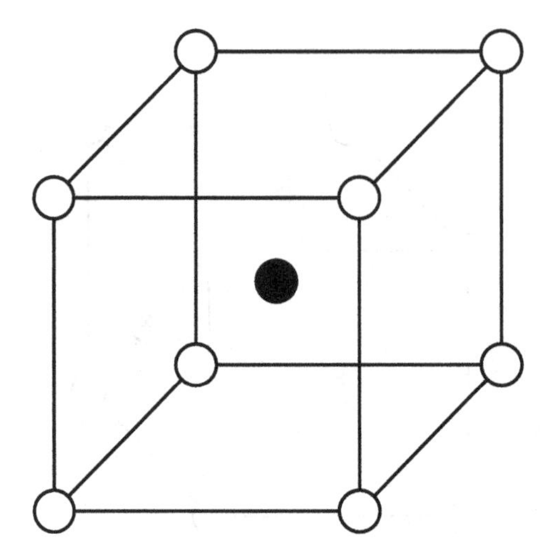

Figure 13.8: The unit body-centered cubic lattice. The cube-edge atoms (∘) form a SC lattice. The body-center atoms (•) form another set of a SC sublattice.

In conclusion, if CR experiments are performed and analyzed properly, then we can obtain detailed information of the Fermi surfaces including the effective mass values.

Dresselhaus, Kip and Kittel mentioned extra CR peaks, which cannot be assigned to the "hole" or "electron" energy bands so far discussed. A FCC lattice can be decomposed into two sublattices (SC and tetragonal). This lattice decomposition has led to light and heavy "holes" in Si. It also implies that there should be light and heavy "electrons" in Si. Light isotropic "electron" is a candidate for the extra peak.

Problem 13.2.1. Derive the secular equation (13.14). Solve this equation and obtain Equation (13.7).

13.3 Cyclotron Resonance in Al

Metals have higher carrier densities and greater effective masses compared with semiconductors. Besides, they reflect the applied radiation at the surface. All of these make it more difficult to observe CR peaks. The difficulties can, however, be overcome by using an *Azbel–Kaner* (AK) geometry [5] in

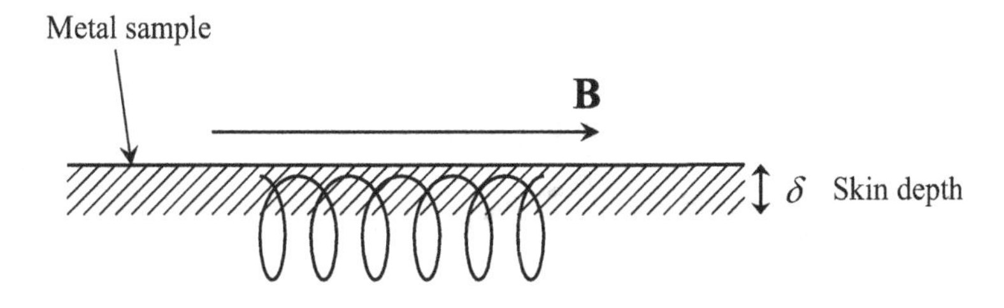

Figure 13.9: Azbel–Kaner geometry. The magnetic field is in the sample surface.

which the magnetic field is applied in the plane of a well-prepared sample surface (see Figure 13.9). The circulating electrons may be probed by the microwave applied in the plane, which penetrates a little within the *skin depth* δ.

Aluminum (Al) forms a FCC lattice, and it is a trivalent metal. Spong and Kip [6] studied the CR in the AK geometry. The most prominent set of anisotropic peaks observed arise from the CPs in {110} planes. In Figure 13.10 the data (dots) and the theoretical curves (solid lines) based on Shockley's formula: [7]

$$\frac{\omega}{eB} = \left(\frac{m_2 m_3 \cos^2(\mu, x_1) + m_3 m_1 \cos^2(\mu, x_2) + m_1 m_2 \cos^2(\mu, x_3)}{m_1 m_2 m_3} \right)^{1/2} \tag{13.19}$$

with the effective masses

$$(m_1, m_2, m_3) = (0.108, 0.156, 1.96)m \quad \text{for Al [110] ellipsoid} \tag{13.20}$$

are shown. Note the excellent fits above $\omega/\omega_{[100]} = 0.5$, where the CR peaks are sharp, and can be clearly identified.

Shockley's formula (13.19) can be derived as follows. After substituting $v_j(t) = \exp(-i\omega t)v_j$ into

$$m_j \frac{dv_j}{dt} = q(\mathbf{v} \times \mathbf{B})_j, \qquad j = 1, 2, 3, \tag{13.21}$$

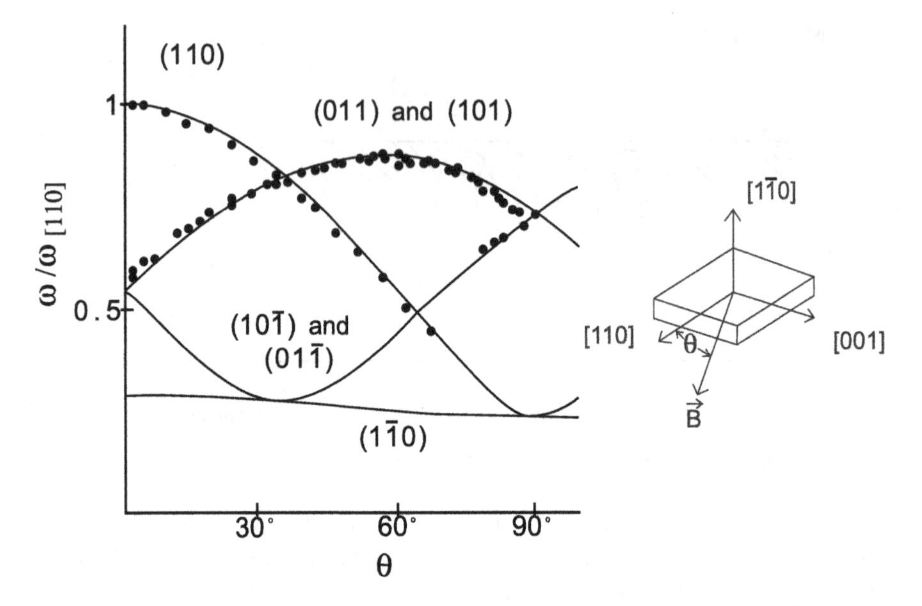

Figure 13.10: CR frequency of Al [110] cylinders data (dots) after Spong and Kip [6] and theoretical curves (solid lines) based on Shockley's formula, Equation (13.19), with $(m_1, m_2, m_3) = (0.108, 0.156, 1.96)m$. The field angle θ is measured from [110] direction in the plane as shown in the inset.

we obtain the secular equation from the condition that the velocity amplitudes (v_1, v_2, v_3) are not identically zero. Solving this secular equation, we obtain the desired expression [Equation (13.19)] for the cyclotron frequency ω (Problem 13.3.1). The obtained frequency ω does not depend on the circulating speed, and hence, it does not depend on the electron's energy. This means that all the electrons resonate at the same frequency when the energy-momentum relation is quadratic; the observed peak should, therefore, be strong and sharp in the absence of scatterers (impurities, phonons).

By assuming $m_1 = m_2$, Equation (13.19) is reduced to DKK formula (13.7). If $m_3 = \infty$, formula (13.19) is reduced to a *cosine law* formula:

$$\omega = eB(m_1 m_2)^{-1/2} \cos\theta, \tag{13.22}$$

where θ is the angle between the field and the c-axis. An example of this case occurs in layered conductors (2D system) (see Section 11.4).

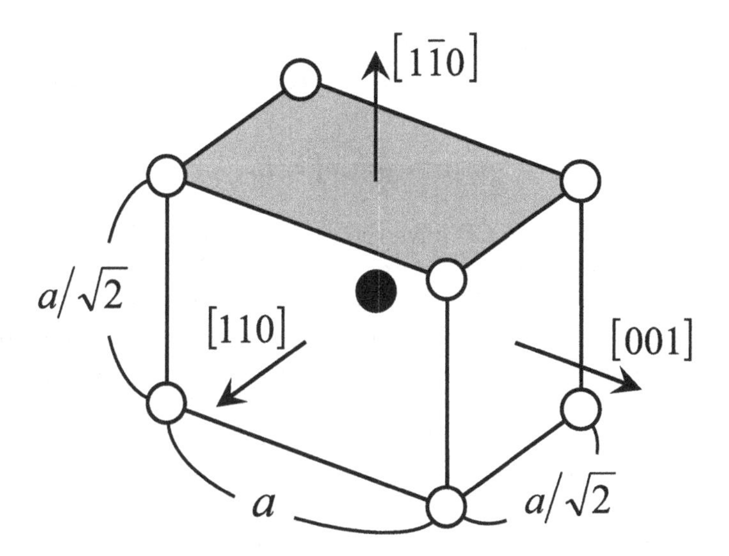

Figure 13.11: The CP (1$\bar{1}$0) in which the electron completes a closed orbit contains periodic arrays of Al (open circles) forming a rectangle of side lengths a and $a/\sqrt{2}$. The adjacent CP are separated by $a/\sqrt{2}$. The tetragonal unit contains Al (closed circle) at the body center.

First, we explain why the electron moves in a three-effective-mass environment. Start with the FCC unit lattice shown in Figure 13.7 (a). Proceeding in the [1$\bar{1}$0] direction we recognize CPs spaced by $a/\sqrt{2}$; each CP contains Al forming a unit rectangle with side length a and $a/\sqrt{2}$ as shown in Figure 13.11. Because the side lengths are different, the electron should move anisotropically in this plane with two different masses (m_1, m_2), $m_1 \neq m_2$. The electron may move with a different mass m_3 in the third direction since the body-center atoms, marked by the blackened circles $\bullet$, impede the motion in the [1$\bar{1}$0] direction although the separation between the adjacent CPs is $a/\sqrt{2}$ and equal to one of the side lengths of the rectangle. In Figure 13.11 open and closed circles represent the same Al.

We plotted the reduced CR frequency ω/ω_{100} as a function of θ in Figure 13.10. This representation is more convenient. In particular the highest CR frequency appears at [110] direction, which is perpendicular to the CP (110). A further practical advantage of this representation will be expounded below.

We now discuss the set of the effective masses obtained. The cyclotronic motion in the (x_1, x_2) plane is found to be characterized by masses (m_1, m_2) = (0.108, 0.156)m, meaning that "electrons" move around the constant-energy orbit:

$$\frac{1}{2}m_1 v_1^2 + \frac{1}{2}m_2 v_2^2 = E_c. \tag{13.23}$$

The effective masses in the CP have the same sign since the curvature signs must be the same so that the electron executes a closed orbit. We assume here that the Al orbits are "electron"-like. The inter ionic distance is shorter by the factor $\sqrt{2}$ in the x_1-direction [110] than in the x_2-direction [001] (see Figure 13.11). The motion in the x_1-direction is more mobile and is characterized by the lighter mass. In the third direction $[1\bar{1}0]$ the electron motion is characterized by $m_3 = 1.96m$. This value is reasonable since the motion is impeded by the layers of the body-center ions as mentiond earlier.

Problem 13.3.1. Derive Shockley's formula, Equation (13.19).

13.4 Cyclotron Resonance in Pb

Khaikin and Minas [8] and Onuki, Suematsu, and Tamura [9] made extensive CR studies in lead (Pb), using the AK geometry. Lead, which forms an FCC crystal, has complicated Fermi surfaces as mentioned earlier in Section 9.2. Figure 13.12 represents the data (circles) of the CR peaks ($\omega \equiv \omega_m$) in the $(1\bar{1}2)$ sample plane, reproduced after Onuki, Suematsu, and Tamura [9]. The θ is the field angle in the $(1\bar{1}2)$ AK plane measured from the axis [110] as shown in the inset. We took the reduced CR frequency ω/ω_0 as the ordinate in place of the apparent effective mass m^*. (The advantages of our representation will be explained later.) Theoretical solid curves [10] are obtained in terms of Shockley's formula, Equation (13.19), with the effective masses

$$(m_1,\, m_2,\, m_3) = (0.244,\ 1.18,\ -8.71)m \tag{13.24}$$

along the axes: $(x_1, x_2, x_3) = ([110], [001], [1\bar{1}0])$. As we can see in Figure 13.12, the fits between theory and experiment are quite good especially for those data at the high frequencies (above $\omega/\omega_0 > 0.6$).

Onuki, Suematsu, and Tamura [9] observed the angular dependence of the CR peaks in other geometries including those for the $(1\bar{1}0)$, (100), and (111)

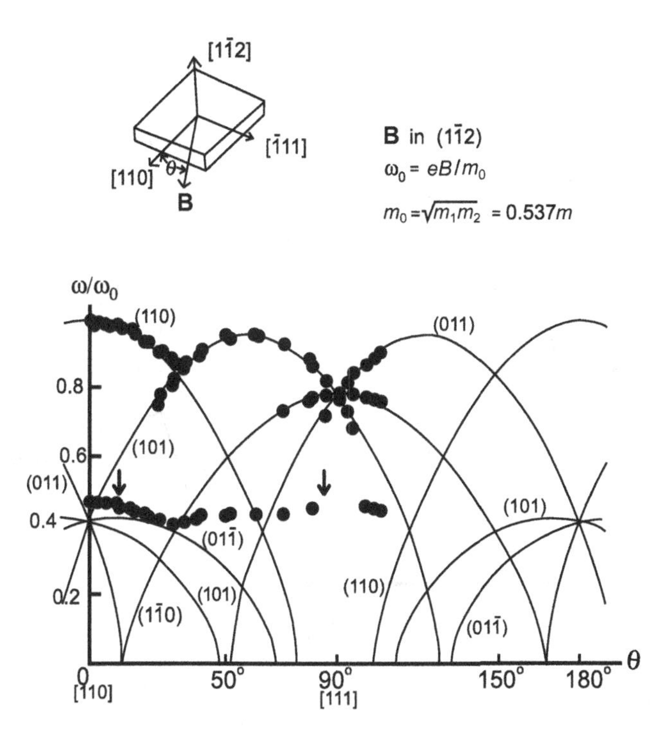

Figure 13.12: Experimental data (circles) of the CR peaks in the $(1\bar{1}2)$ sample plane are reproduced after Onuki, Suematsu, and Tamura [9]. The angle θ denotes the field angle in the $(1\bar{1}2)$ plane measured from the [110] axis, as shown in the inset. Theoretical solid curves are obtained through Shockley's formula, Equation 13.19, with $(m_1, m_2, m_3) = (1.18, 0.244, -8.71)m$; the six $\{110\}$ planes are taken as the cyclotronic planes.

AK planes. Figure 13.13 represents the data for the $(1\bar{1}0)$ AK plane. Fits between theory and experiments were obtained with the *same* set of (m_1, m_2, m_3). We plotted the reduced CR frequency as a function of θ in Figures 13.12 and 13.13. Analysis in terms of formula (13.19) is more convenient in this representation of the data. The reduced frequency never grows to ∞, while the apparent effective mass can blow up. A further practical advantage will be expounded below.

We now discuss the set of effective masses obtained. The cyclotronic motion in the (x_1, x_2) plane is characterized by the masses (m_1, m_2) of the same sign. We may assign positive values to the pair: $(m_1, m_2) = (0.244,$

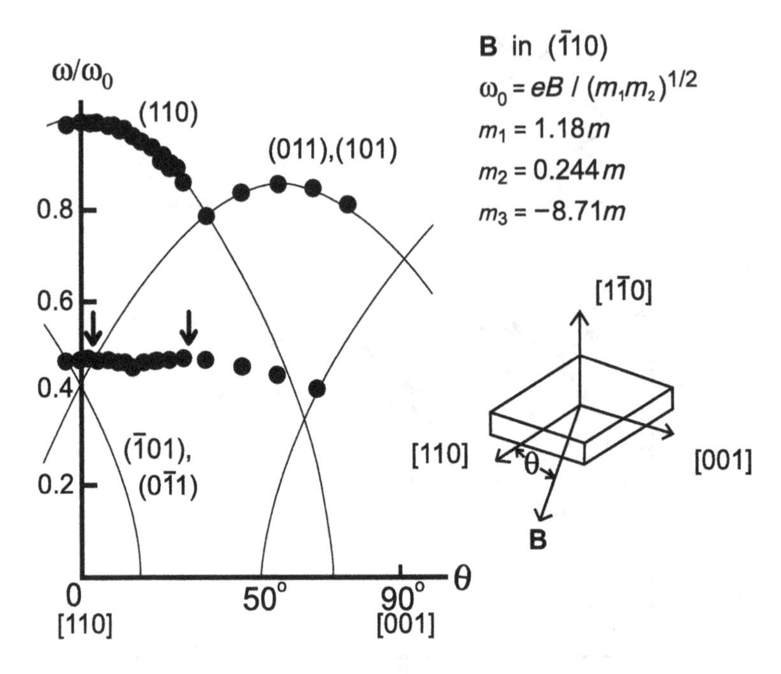

Figure 13.13: The reduced CR peaks for the $(1\bar{1}0)$ sample plane in Pb are plotted as functions of the field angle θ measured from the [110] axis, as shown in the inset. The experimental data were reproduced after Onuki, Suematsu, and Tamura [9] and theoretical curves are obtained from Equation (13.19) with $(m_1, m_2, m_3) = (1.18, 0.244, -8.71)m$ for the cyclotronic planes $\{110\}$.

$1.18)m$, meaning that "electrons" move around the constant-energy orbit:

$$\frac{1}{2}m_1v_1^2 + \frac{1}{2}m_2v_2^2 = E_c. \tag{13.25}$$

The interionic distance is shorter by the factor $\sqrt{2}$ in the x_1-direction than in the x_2-direction. The motion in the x_1-direction is more mobile and is characterized by the smaller effective mass $m_1 = 0.244m$. The motion in the x_3-direction is found to be characterized by effective mass m_3 with a *different sign*: $m_3 = -8.71m$. This means that the cyclotronic motion is "electron"-like while the helical-axial motion is "hole"-like, which appears to be a little strange from the common-sense point of view. But this does happen without violating any physical principles.

To see this, let us recall that the values of $\{m_j\}$ were obtained by fitting the theoretical curves with those data appearing in the high-frequency region ($\omega/\omega_c > 0.6$) in Figure 13.12. Here, the agreement between theory and experiment is well within the measurement errors. Since m_3 has a different sign from that of m_1 (m_2), the CR frequency ω calculated through formula (13.19) vanishes at certain angles. For example, the CR frequency arising from the (110) plane falls down to zero at about 75° and reappears at about 105°; there is no resonance peak between these two angles. It is noted here that the absence of the CR peaks can be discussed most simply in our representation of the reduced CR frequency as the ordinate. Similar features hold for all other five cases originating from the $\{110\}$ family. In particular, the two theoretical curves coming from the (101) and $(01\bar{1})$ planes never rise above 0.42 over the whole angular range (0, 180°); the two curves have their maxima (value 0.48) near $\theta = 0$, and the experimental data support their existence in this neighborhood. In general, the fits between theory and experiment for ω/ω_c below the value 0.5 are only fair, but they are consistent.

Let us now examine the case of the $(\bar{1}10)$ AK plane shown in Figure 13.13. Essentially, the same things can be said about the fits. The data in Pb here are plotted using the same representation and geometry as those in Al shown in Figure 13.10. Hence, there are similarities between the two figures. However, there are no CR peaks for the (110) plane corresponding to the fact that Equation (13.19) gives a pure imaginary number for the whole angular range. Due to the special measurement geometry, the CR peaks are doubly degenerate for the two pairs of CPs: [(011), (101)], $[(\bar{1}01), (0\bar{1}1)]$. In conclusion, the number of the CR peaks *and* the angular dependence of each CR maximum for all experimental geometries can be described by formula (13.19) with $(m_1, m_2, m_3) = (0.244, 1.18, -8.71)m$. This situation parallels the cases of Ge and Si (semiconductors), where the measurments of the CR peaks led to the determination of the effective masses for "electrons" and "holes." The midsections of the cylinders shown in Figure 9.8, are found to be *hyperboloidal*. "Electrons" and "holes" coexist near the "neck."

Onuki, Suematsu, and Tamura [9] observed other CR peaks not belonging to the $\langle 110 \rangle$ family. In fact, we observe that in both Figure 13.12 and Figure 13.13 there are angle-independent peaks (circles) at $\omega/\omega_c = 0.45$, indicated by arrows, with the effective mass:

$$m^* = 1.30m \qquad \text{for the } \langle 100 \rangle \text{ cyclotronic planes.} \tag{13.26}$$

These peaks can be identified with those originating from the CPs in $\langle 100 \rangle$

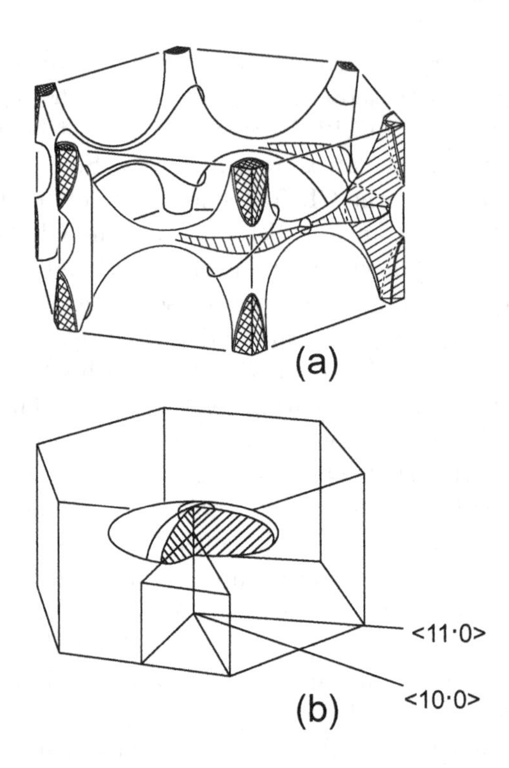

Figure 13.14: The Fermi surface of zinc (Zn): (a) first and second bands (holes) and (b) third band (electrons; the lens is at the center in (b) after Gibbons and Falicov [12]. The Fermi surfaces in cadmium (Cd) are very similar to those in Zn.

with all equal masses $m_1 = m_2 = m_3 = m^*$.

13.5 Cyclotron Resonance in Zn and Cd (HCP)

Zinc (Zn) and cadmium (Cd), both HCP, show similar angular dependence in CR peaks [11, 12]. For both metals, the most visible peaks are those coming from the central orbit on the *lenslike Fermi surface* in the third zone ("electron") [13], shown in Figure 13.14 (b). The data of the CR peaks for the central orbits in the "lens" in Zn and Cd are shown in Figure 13.15, where the reduced effective masses are chosen as the ordinates. By symmetry consideration the motion in the hexagonal plane is characterized by equal masses $m_1 = m_2 = m_t$, and the motion along the c-axis is characterized by a different mass m_l.

For the data obtained by Shaw, Eck, and Zych [11] in Figure 13.15 the CPs are identified as $\{11{\cdot}0\}$ and the CR peaks depend on the field angle θ from the c-axis, following the DKK formula, Equation (13.7). The apparent effective mass m^* is from Equation (13.7):

$$m^* \equiv \frac{eB}{\omega_c} = \frac{1}{[m_t^{-2} \cos^2 \theta + (m_t m_l)^{-1} \sin^2 \theta]^{1/2}}. \tag{13.27}$$

We obtain the best fits as follows.

$$m_t = 1.04m, \quad m_l = 0.212m \quad \text{for Zn lens,} \tag{13.28}$$

$$m_t = 1.14m, \quad m_l = 0.217m \quad \text{for Cd lens.} \tag{13.29}$$

We note that good fits are obtained for all angles. The previously reported fits (dotted lines) [11] based on the theory utilizing the nearly free-electron-model formula

$$\frac{m^*}{m_0} = \left(\frac{2}{\pi}\right) \cos^{-1} \left[\frac{\sin \theta}{(\beta^2 - \cos^2 \theta)^{1/2}}\right], \qquad m_0 \text{ and } \beta = \text{ parameters,} \tag{13.30}$$

deviate significantly from experimental data in the neighborhood of $\theta = 0$ (c-axis), which is shown in Figure 13.15. Our fits are distinctly better.

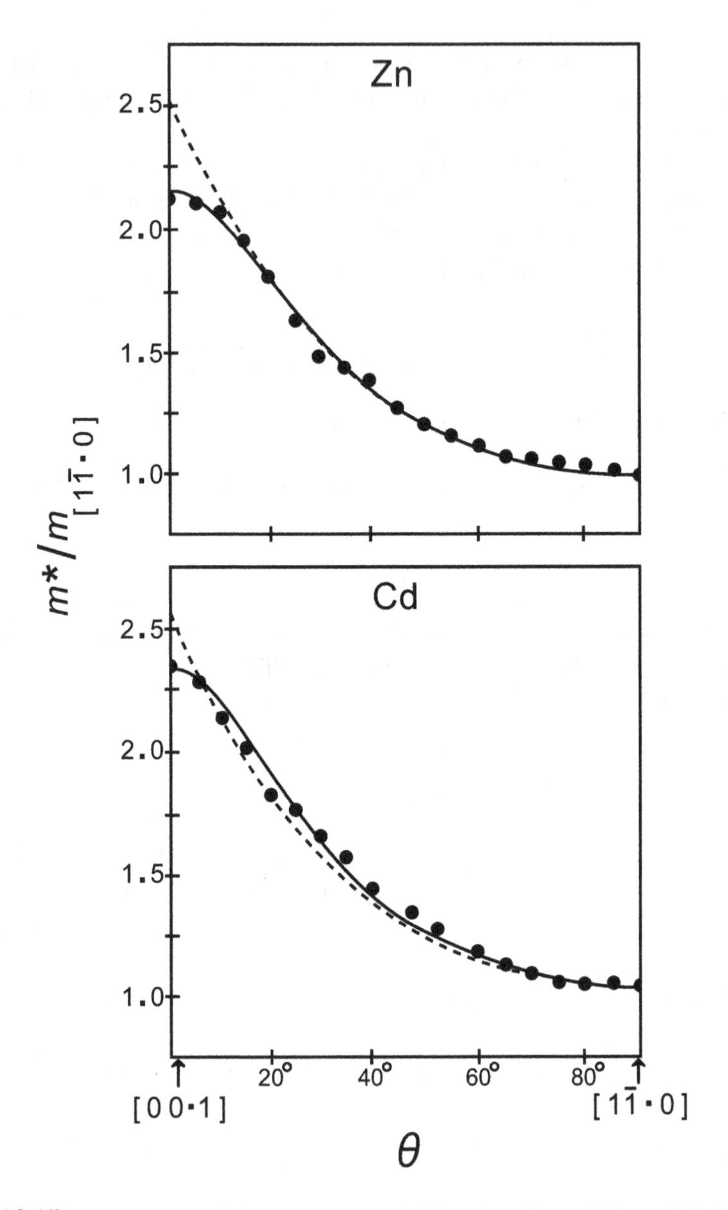

Figure 13.15: Comparison of the experimental data after Shaw, Eck, and Zych. (dots) [11] and theories for the central orbits on the lens in Zn and Cd (HCP). Solid lines are calculated based on Equation (13.7); dashed lines are based on the nearly-free-electron model (data and dashed lines are from Figure 6 in Ref. 11).

Chapter 14

Seebeck Coefficient (Thermopower)

Based on the idea that different temperatures generate different carrier densities and the resulting carrier diffusion causes the thermal electromotive force (emf), a new formula for the Seebeck coefficient (thermopower) S is obtained: $S = (2\ln 2/3)(qn)^{-1}\epsilon_F k_B(\mathcal{N}_0/V)$, where q, n, ϵ_F, $\mathcal{N}_0$, and V are charge, carrier density, Fermi energy, density of states at ϵ_F, and volume, respectively. Ohmic and Seebeck currents are fundamentally different in nature, and hence, cause significantly different transport behaviors. For example, the Seebeck coefficient S in Cu is positive, while the Hall coefficient is negative. In general, the Einstein relation between the conductivity and the diffusion coefficient does not hold for a multicarrier metal.

14.1 Introduction

When a metallic bar is subjected to a voltage (V) or a temperature (T) difference, an electric current is generated. For small voltage and temperature gradients we may assume a linear relation between the electric current density $\mathbf{j}$ and the gradients:

$$\mathbf{j} = \sigma(-\nabla V) + A(-\nabla T) = \sigma\mathbf{E} - A\nabla T, \tag{14.1}$$

where σ is the conductivity. If the ends of the conducting bar are maintained at different temperatures, no electric current flows. Thus, from Equation (14.1),

$$\sigma\mathbf{E}_s - A\nabla T = 0, \tag{14.2}$$

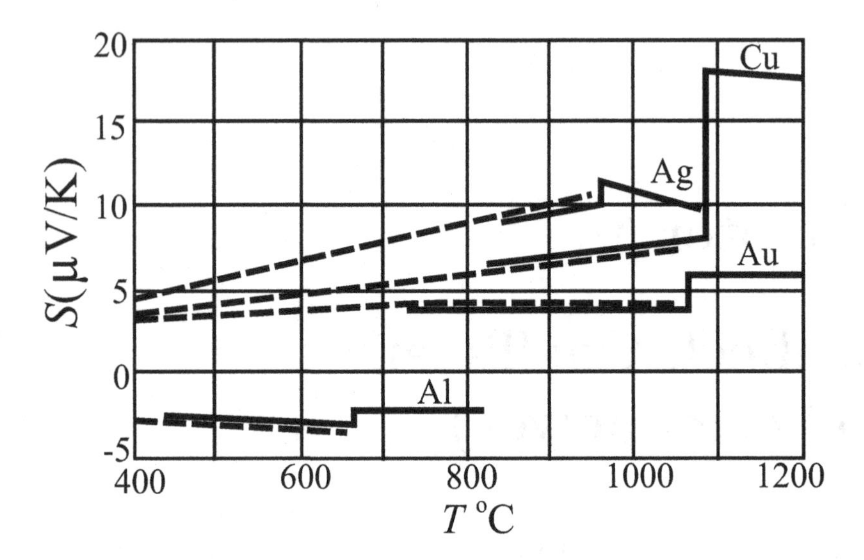

Figure 14.1: High temperature Seebeck coefficients above 400°C for Ag, Al, Au, and Cu. The solid and dashed lines represent two experimental data sets.

where $\mathbf{E}_s$ is the field generated by the thermal electromotive force (emf). The *Seebeck coefficient* (*thermopower*) S is defined through

$$\mathbf{E}_s = S\nabla T, \qquad S \equiv \frac{A}{\sigma}. \tag{14.3}$$

The conductivity σ is positive, but the Seebeck coefficient S can be positive or negative. As shown in Figure 14.1, the measured Seebeck coefficient S in Al at high temperatures (400–700°C) is negative, while the S in noble metals (Cu, Ag, Au) are positive (e.g. Ref.1).

Based on the classical idea that different temperatures generate different electron drift velocities, we obtain (Problem 14.1.1)

$$S = -\frac{c_v}{3ne}, \tag{14.4}$$

where c_v is the specific heat. Setting c_v equal to $3nk_B/2$, we obtain the *classical formula* for thermopower:

$$S_{\text{classical}} = -\frac{k_B}{2e} = -0.43 \times 10^{-4} \ \text{ V K}^{-1}. \tag{14.5}$$

Observed metallic Seebeck coefficients at room temperature are of the order of microvolts per degree (see Figure 14.1), a factor of 10 smaller than $S_{\text{classical}}$. If we introduce the Fermistatistically computed specific heat $c_v = (\pi^2/3)k_B^2 T \mathcal{N}_0$ in Equation (14.4), we obtain

$$S_{\text{semiquantum}} = -\frac{\pi}{6}\frac{k_B}{e}\left(\frac{k_B T}{\epsilon_F}\right), \tag{14.6}$$

which is often quoted in materials handbooks [1]. Formula (14.6) remedies the difficulty with respect to magnitude. But the correct theory must explain the two possible signs of S besides the magnitude.

Fujita, Ho, and Okamura [2] developed a quantum theory of the Seebeck coefficient. We follow this theory, explaining the sign and the T-dependence of the Seebeck coefficient.

Problem 14.1.1. Derive formula (14.4) using kinetic theory.

14.2 Quantum Theory

We assume that the carriers are conduction electrons, each having charge q [$-e$ for "electron" ($+e$ for "hole")] and effective mass m^*. Assuming a one-component system, the conductivity σ after standard kinetic theory is

$$\sigma = \frac{q^2 n \tau}{m^*}, \tag{14.7}$$

where n is the carrier density and τ the mean free time. Note that σ is always positive irrespective of whether $q = -e$ or $+e$. The Fermi distribution function f is

$$f(\epsilon;\, \beta,\, \mu) = \frac{1}{e^{\beta(\epsilon-\mu)}+1}, \qquad \beta \equiv (k_B T)^{-1}, \tag{14.8}$$

where μ is the chemical potential whose value at 0 K equals the Fermi energy ϵ_F. The voltage difference $\Delta V = LE$, with L being the sample length, generates the chemical potential difference $\Delta\mu$, the change in f, and consequently, the electric current. Similarly, the temperature difference ΔT generates the change in f and the current.

We assume a high Fermi degeneracy: $T_F \gg T$. At 0 K the Fermi surface is sharp and there are no conduction electrons. At a finite T, "electrons"

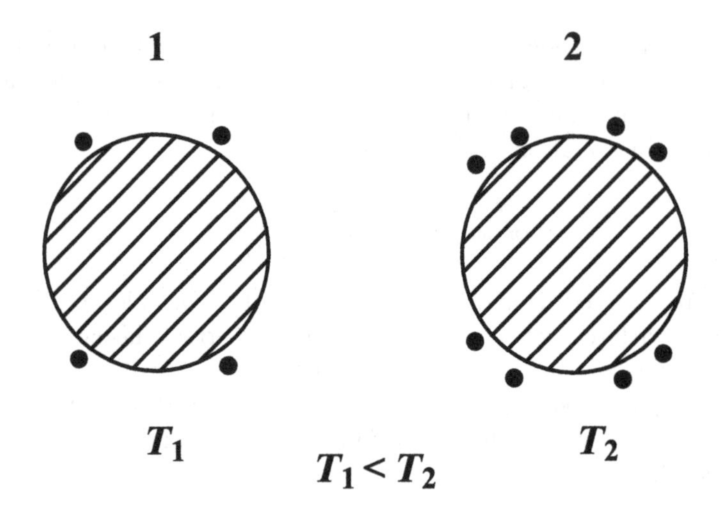

Figure 14.2: More "electrons" are excited at the high temperature end: $T_2 > T_1$. "Electrons" diffuse from 2 to 1.

("holes") are thermally excited near the Fermi surface if the curvature of the surface is negative (positive) see (Figures. 14.2 and 14.3). Consider first the case of "electrons." The number of thermally excited "electrons," N_x, having energies greater than the Fermi energy ϵ_F is defined and calculated as [2] (Problem 14.2.1)

$$N_x \equiv \int_{\epsilon_F}^{\infty} d\epsilon \, \mathcal{N}(\epsilon) \frac{1}{e^{\beta(\epsilon-\mu)}+1} = \ln 2 \; k_B T \mathcal{N}_0, \quad \mathcal{N}_0 = \mathcal{N}(\epsilon_F), \tag{14.9}$$

where $\mathcal{N}$ is the density of states. The excited particle density $n \equiv N_x/V$ is higher at the high-temperature end, and the particle current runs from the high- to the low-temperature end. This means that the electric current runs towards (away from) the high-temperature end in an "electron" ("hole")-rich material. After using Equations (14.1) and (14.3), we obtain

$$S < 0 \qquad \text{for "electrons,"}$$

$$S > 0 \qquad \text{for "holes."} \tag{14.10}$$

The Seebeck current arises from the thermal diffusion. We assume Fick's

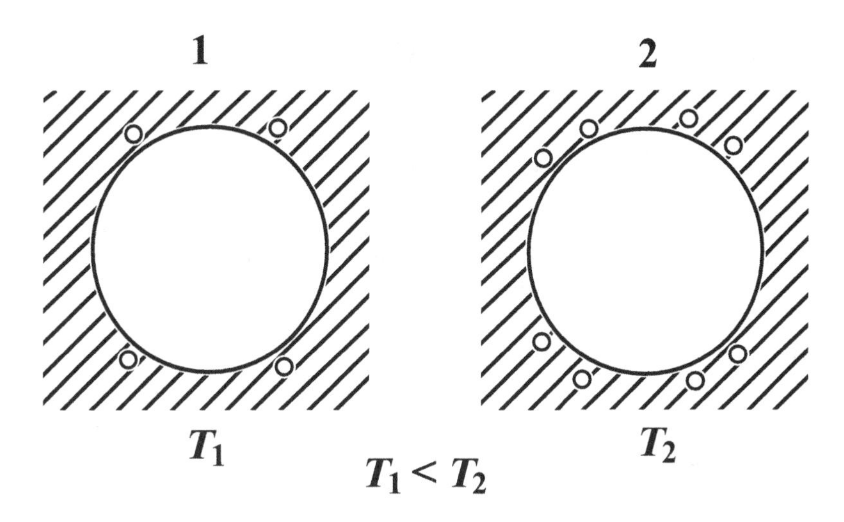

Figure 14.3: More "holes" are excited at the high temperature end: $T_2 > T_1$. "Holes" diffuse from 2 to 1.

law:

$$\mathbf{j} = q\,\mathbf{j}_{\text{particle}} = -qD\nabla n, \tag{14.11}$$

where D is the diffusion coefficient, which is computed from the standard formula

$$D = \frac{1}{d}vl = \frac{1}{d}v_F^2\tau, \qquad v = v_F, \tag{14.12}$$

where d is the dimension. (In this chapter the dimension is denoted by d instead of D, which denotes the diffusion coefficient.) The density gradient ∇n is generated by the temperature gradient ∇T and is given by

$$\nabla n = \frac{\ln 2}{Vd}k_B\mathcal{N}_0\nabla T, \tag{14.13}$$

where Equation (14.9) is used. Using the last three equations and Equation (14.1), we obtain

$$A = \frac{\ln 2}{V}qv_F^2k_B\mathcal{N}_0\tau. \tag{14.14}$$

Using Equations (14.3), (14.7), and (14.14), we obtain

$$\boxed{S = \frac{A}{\sigma} = \frac{2\ln 2}{d}\left(\frac{1}{qn}\right)\epsilon_F k_B \frac{\mathcal{N}}{V}.} \tag{14.15}$$

The relaxation time τ is not present in this expression because it cancels out from the numerator and the denominator.

Problem 14.2.1. Verify Equation (14.9).

14.3 Discussion

The derivation of our formula, Equation (14.15), for the Seebeck coefficient S was based on the idea that the Seebeck emf arises from thermal diffusion. We used the high Fermi degeneracy condition: $T_F \gg T$. The relative errors due to this approximation *and* due to the neglect of the T dependence of μ are both of the order $(k_B T/\epsilon_F)^2$. Formula (14.15) can be negative or positive, while the materials handbook formula (14.6) has the negative sign. The average speed v for highly degenerate electrons is equal to the Fermi velocity v_F (independent of T). Hence, semi classical Equations (14.4) through (14.6) break down. In Ashcroft and Mermin's (AM) book [3] the origin of a positive S in terms a mass tensor $M = \{m_{ij}\}$ is discussed. This tensor M is real and symmetric, and hence, it can be characterized by the principal masses $\{m_j\}$. Formula for S obtained by AM [Equation (13.62) in Ref. 3], can be positive or negative but is hard to apply in practice. In contrast our formula (14.15) is interpreted straightforwardly. Besides our formula for a one-carrier system is T independent, while the AM formula is linear in T. This difference arises from the fact that the thermal diffusion is the cause of the Seebeck current.

Formula (14.15) is remarkably similar to the standard formula for the Hall coefficient:

$$R_H = (qn)^{-1}. \tag{14.16}$$

Both Seebeck and Hall coefficients are inversely proportional to charge q, and hence, they give important information about the carrier charge sign. In fact the measurement of the thermopower of a semiconductor can be used to see if the conductor is n-type or p-type (with no magnetic measurements).

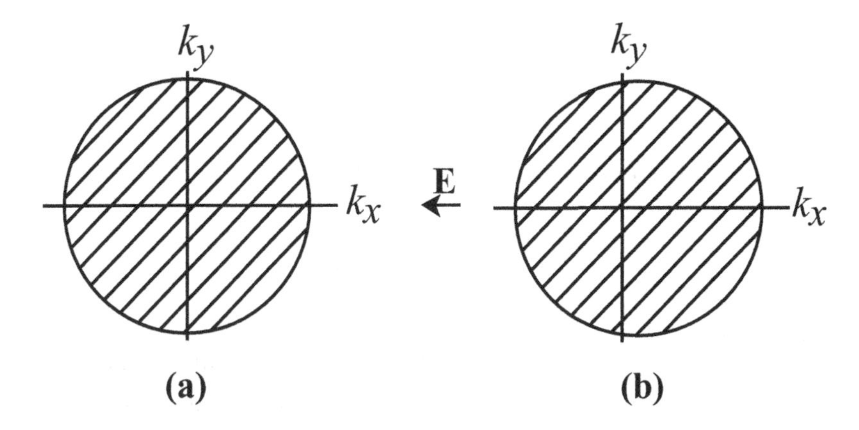

Figure 14.4: Due to the electric field $\mathbf{E}$ pointed in the negative x-direction, the steady-state electron distribution in (b) is generated, which is a translation of the equilibrium distribution in (a) by the amount $\hbar^{-1}eE\tau$.

If only one kind of carrier exists in a conductor, then the Seebeck and Hall coefficients must have the same sign as observed in alkali metals.

Let us consider the electric current caused by a voltage difference. The current is generated by the electric force that acts on *all* electrons. The electron's response depends on its mass m^*. The density (n) dependence of σ can be understood by examining the current-carrying steady state in Figure 14.4 (b). The electric field $\mathbf{E}$ displaces the electron distribution by a small amount $\hbar^{-1}qE\tau$ from the equilibrium distribution in Figure 14.4 (a). Since all the conduction electrons are displaced, the conductivity σ depends on the particle density n. The Seebeck current is caused by the density difference in the thermally excited electrons near the Fermi surface, and hence, the thermal diffusion coefficient A depends on the density of states at the Fermi energy $\mathcal{N}_0$ [see Equation (14.14)]. We further note that the diffusion coefficient D does not depend on m^* directly [see Equation (14.12)]. Thus, the Ohmic and Seebeck currents are fundamentally different in nature.

For a single-carrier metal such as the alkali metal Na, where only "electrons" exist, both R_H and S are negative. The *Einstein relation* between the conductivity σ and the diffusion coefficient D holds:

$$\sigma \propto D. \tag{14.17}$$

Using Equations (14.7) and (14.12), we obtain

$$\frac{D}{\sigma} = \frac{v_F^2 \tau/3}{q^2 n\tau/m^*} = \frac{2}{3}\frac{\epsilon_F}{q^2 n}, \tag{14.18}$$

which is a material constant. The Einstein relation is valid for a singlecarrier system. But the relation does not hold in general for multicarrier systems. The ratio D/σ for a two-carrier system containing (1) "electrons" and (2) "holes" is given by

$$\frac{D}{\sigma} = \frac{(1/3)v_1^2\tau_1 + (1/3)v_2^2\tau_2}{q_1^2(n_1/m_1)\tau_1 + q_2^2(n_2/m_2)\tau_2}, \tag{14.19}$$

which is a complicated function of (m_1/m_2), (n_1/n_2), (v_1/v_2), and (τ_1/τ_2). In particular the mass ratio m_1/m_2 may vary significantly for a heavy fermion condition, which occurs whenever the Fermi surface just touches the Brillouin boundary, see below. An experimental check on the violation of the Einstein relation can be carried out by simply examining the T dependence of the ratio D/σ. This ratio from Equation (14.18) is constant for a single-carrier system, while from Equation (14.19) it depends on T since the generally T-dependent mean free times (τ_1, τ_2) arising from the electron-phonon scattering do not cancel out from numerator and denominator. Conversely, if the Einstein relation holds for a metal, the spherical Fermi surface approximation with a single effective mass m^* is valid for this single-carrier metal.

Figure 14.1 shows the T dependence of the Seebeck coefficient S for selected metals. Note that the S for the familiar noble metals (Cu, Ag, Au) is positive, and each curve shows a weak T dependence above 100 K. In particular the S for Au is positive and nearly T independent up to the melting point (1060°C), which cannot be explained in terms of the T-linear semiquantum formula (14.6). The Hall coefficient R_H for the noble metals is negative. The reason why S and R_H have opposite signs must be explained.

Equation (14.15) with $d = 3$ is

$$S = \frac{2\ln 2}{3}\frac{1}{qn}\epsilon_F k_B \frac{\mathcal{N}_0}{V}. \tag{14.20}$$

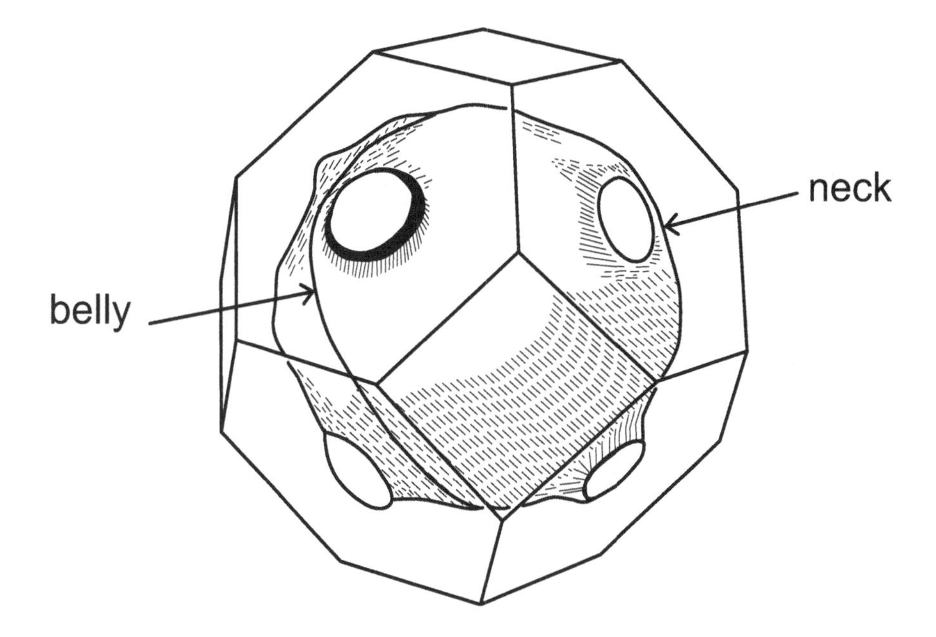

Figure 14.5: The Fermi surface of silver (FCC) has "necks", with the axes in the ⟨111⟩ direction, located near the Brillouin boundary.

The Fermi surface in Ag is far from spherical and has a set of "necks" at the Brillouin boundary (see Figure 14.5) [4]. The curvatures along the axes of each neck are positive, and hence, the Fermi surface is "hole"-generating. Experiments indicate [4] that the minimum neck area A_{111}(neck) in the k-space is 1/51 of the maximum belly area A_{111}(belly), meaning that the Fermi surface just touches the Brillouin boundary (Figure 14.5 exaggerates the neck area). The density of "hole"-like states, $\mathcal{N}_{\text{hole}}$, associated with the ⟨111⟩ necks, having the heavy-fermion character due to the rapidly varying surface with energy, is much greater than that of "electron"-like states, $\mathcal{N}_{\text{electron}}$, associated with the ⟨100⟩ belly. The thermally excited "hole" density is higher than the "electron" density, yielding a positive S. The principal mass m_1 along the axis of a small neck ($m_1^{-1} = \partial^2\epsilon/\partial p_1^2$) is positive ("hole"-like) and extremely large. The "hole" contribution to the conduction is small ($\sigma \propto m^{*-1}$), as is the "hole" contribution to the Hall voltage. Then the "electrons" associated with the nonneck Fermi surface dominate and yield a negative Hall coefficient R_H.

Formula (14.15) indicates that thermal diffusion contribution to S is T independent. The observed S in many metals is mildly T dependent. For example, the coefficient S for Ag increases slightly before melting, while the coefficient S for Au is nearly constant and decreases. These behaviors arise from the incomplete compensation of the scattering effects. "Electrons" and "holes" that are generated from the complicated Fermi surfaces will have different effective masses and densities, and the resulting incomplete compensation of τ's yields a T dependence.

Chapter 15

Infrared Hall Effect

If a linearly polarized laser is applied to a metal sample in the Faraday geometry, the transmitted radiation becomes elliptically polarized with the axis rotated by the Faraday angle θ_F. This angle $\theta_F(\omega, T)$ is a function of the laser frequency ω and the temperature T. The small frequency-dependent (dynamic) θ_F is proportional to the dynamic Hall angle $\theta_H(\omega, T)$. A kinetic theory for the infrared (IR) Hall effect is developed in this chapter. The ratio $\mathrm{Re}[\cot\theta_H]/\mathrm{Im}[\cot\theta_H] = -\gamma_H(\omega, T)/\omega$ *directly* yields the dynamic Hall relaxation rate γ_H as a function of (ω, T). The IR Hall effect measurements in Au and Cu give a remarkable result: $\gamma_H(\omega, T) = \gamma_H(0, T)$. That is, the dynamic rate up to mid IR $\sim 1000\ \mathrm{cm}^{-1}$ is equal to the static rate.

15.1 Introduction

In 2000, Cerne et al. [1] measured IR ($\sim 1000\ \mathrm{cm}^{-1}$) *Faraday rotation angle* θ_F and *circular dichroism* (*optical ellipticity*) in Au (Cu) thin films, using sensitive polarization modulation techniques in the temperature range of 20 K $< T <$ 300 K and magnetic fields up to 8 T. Figure 15.1 shows a schematic sketch of the experimental setup (Faraday geometry). A linearly polarized laser is applied along the magnetic-pole axis. The laser analyzed at the analyzer A after passing the material sample M shows an elliptic polarization accompanied by its major axis rotation. In the measurements the Faraday rotation and the optical ellipticity, both of which are proportional to the sample thickness and the magnetic field, are very small ($\sim 10^{-3}$) (see Figure 15.2).

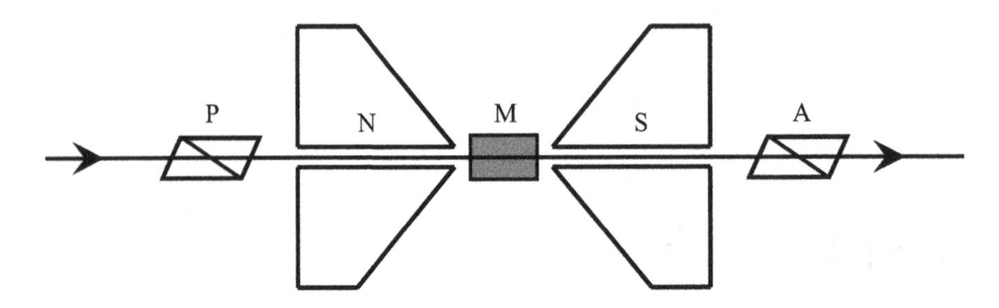

Figure 15.1: A linearly polarized laser beam is sent along the magnetic polar axis in a Faraday geometry. After passing the material sample M, the beam at the analyzer A shows an elliptic polarization with the major axis rotation.

The Faraday angle θ_F is linearly related to the Hall angle θ_H. The Hall angle θ_H is defined such that $E_H = E \tan\theta_H$ (see Figure 15.3). Combined with the zero-field conductivity data, Cerne et al. obtained the real (Re) and imaginary (Im) parts of the dynamic $\cot\theta_H$. In the measurements, the Hall angle θ_H is very small, so that $\tan\theta_H \simeq \theta_H$. The data for temperature (T) dependent $\mathrm{Re}[\theta_H^{-1}]$ and $\mathrm{Im}[\theta_H^{-1}]$ at 1079 cm^{-1} and 8 T in Au are reproduced in Figure 15.4 (a) and (b). We observe that $-$IR $\mathrm{Re}[\theta_H^{-1}]$ (filled circle •) is T-linear and IR $\mathrm{Im}[\theta_H^{-1}]$ (open circle ∘) is positive and ω-independent, $\omega =$ laser frequency:

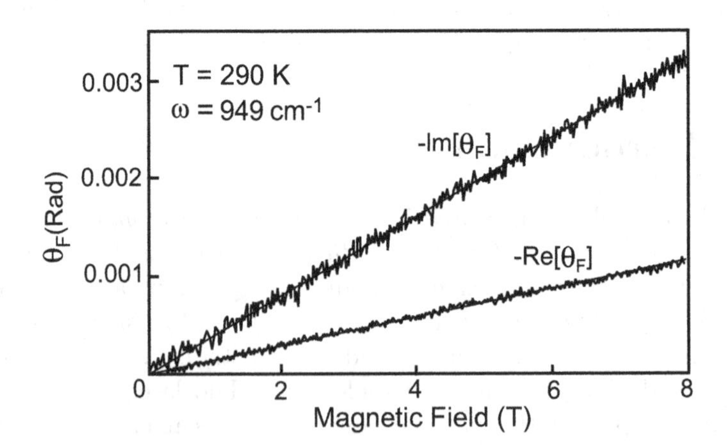

Figure 15.2: Real (Re) and imaginary (Im) parts of the Faraday angle θ_F as a function of the magnetic field for gold after Cerne et al. [1].

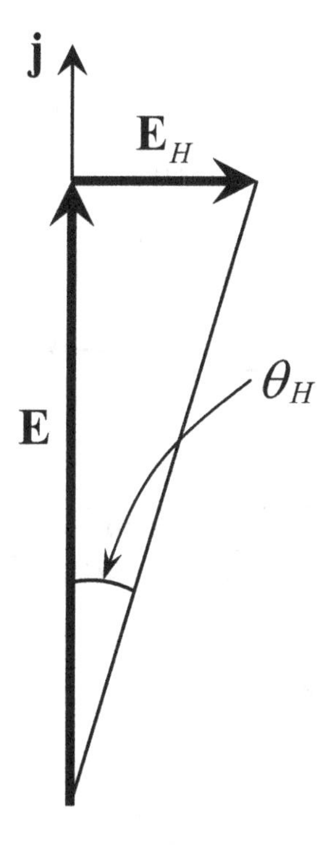

Figure 15.3: The Hall angle θ_H.

$$-\mathrm{Re}[\theta_H^{-1}(\omega,T)] \;\propto\; T,$$

$$\mathrm{Im}[\theta_H^{-1}(\omega,T)] \;=\; T\text{-independent}. \tag{15.1}$$

In Figure 15.4 (b) we see that IR $\mathrm{Im}[\theta_H^{-1}]$ is ω-linear and $-\mathrm{Re}[\theta_H^{-1}]$ is ω-independent:

$$-\mathrm{Re}[\theta_H^{-1}(\omega,T)] \;=\; \omega\text{-independent},$$

$$\mathrm{Im}[\theta_H^{-1}(\omega,T)] \;\propto\; \omega. \tag{15.2}$$

We shall develop a kinetic theory to explain these behaviors in the next section.

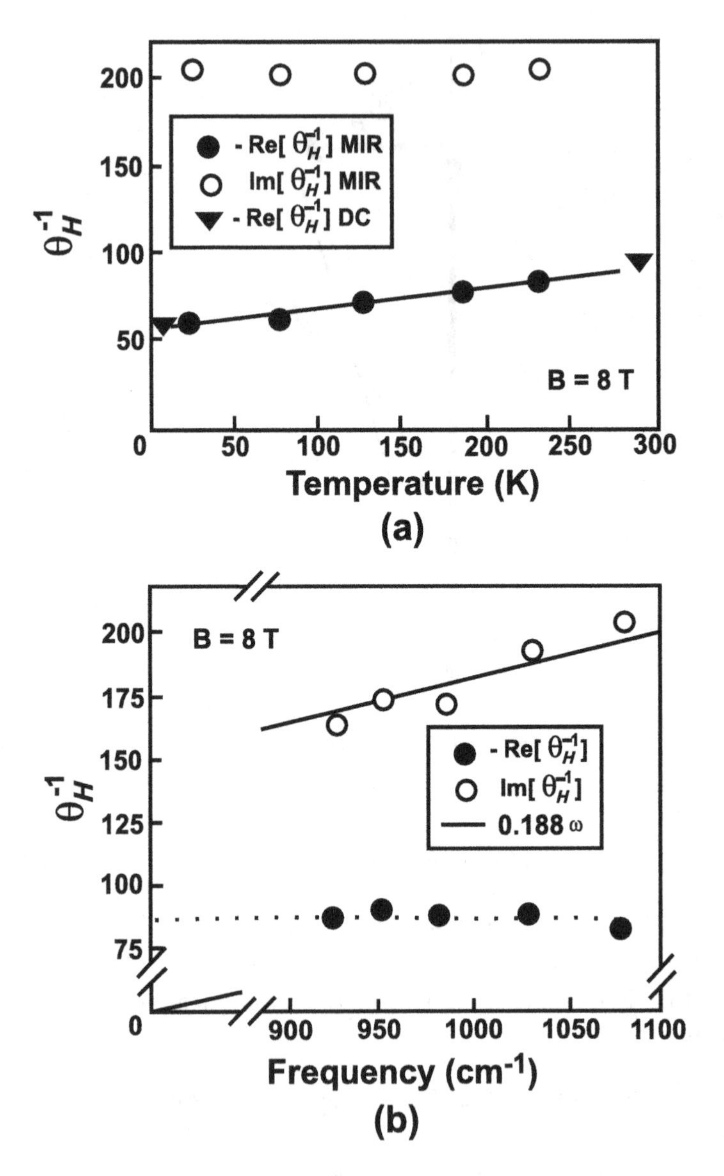

Figure 15.4: The dynamic Hall angle $\theta_H^{-1}(\omega, T)$ at magnetic field of 8 T in Au after Cerne et al. [1]. (a) The temperature dependence at 1079 cm^{-1}. (b) The frequency dependence at 290 K.

15.2 Kinetic Theory

A kinetic theory for the IR Hall effect was developed by Fujita, Kim, and Okamura [2]. We follow this theory.

15.2.1 Conductivity

We first consider the static transport. Assume a weak constant electric field $\mathbf{E}$ applied along the z-axis in a sample containing conduction electrons. Newton's equation of motion with the neglect of the scattering is $m^*(dv_z/dt) = qE$. Solving this equation for v_z and assuming that the acceleration persists in the mean free time τ, we obtain

$$\text{Drift velocity} = v_d = \frac{qE}{m^*}\tau\,. \tag{15.3}$$

The electric current density (z-component) j is

$$j = qnv_d = q^2 n (m^*)^{-1} \tau E\,, \tag{15.4}$$

where n is the electron density. Assuming Ohm's law, $j = \sigma E$, we obtain an expression for the conductivity:

$$\sigma = q^2 n (m^*)^{-1} \gamma_0^{-1}\,, \qquad \gamma_0 \equiv \tau^{-1}\,, \tag{15.5}$$

where $\gamma_0(T)$ is the *static* relaxation rate.

15.2.2 Hall Coefficient

We take a rectangular sample having only "holes" (see Figure 15.5). The current $\mathbf{j}$ runs in the z-direction. Experiments show that if a static magnetic field $\mathbf{B}$ is applied in the y-direction, the sample develops a Hall electric field $\mathbf{E}_H$ so that the Lorentz force be balanced out:

$$e(\mathbf{E}_H + \mathbf{v}_d \times \mathbf{B}) = 0 \qquad \text{or} \qquad E_H(\text{magnitude}) = v_d B\,. \tag{15.6}$$

In the present geometry the field $\mathbf{E}_H$ points in the positive x-direction. We take the convention that E_H is measured relative to this direction, whereby the field E_H is positive (negative) for the "hole" ("electron"). The Hall coefficient R_H is defined and calculated as

$$R_H \equiv \frac{E_H}{jB} = \frac{1}{qn}\,. \tag{15.7}$$

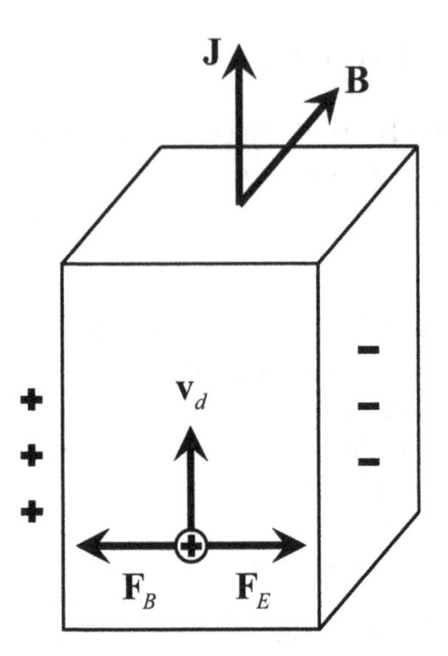

Figure 15.5: The magnetic and electric forces $(\mathbf{F}_B, \mathbf{F}_E)$ balance out to zero in the Hall effect measurement. The carriers are "holes" by assumption.

We note that the Hall coefficient R_H depends on the charge q and the density n. The charge sign is given by Equation (15.7) with the stated convention.

15.2.3 Hall Angle

The *Hall angle* θ_H is the angle between the current $\mathbf{j}$ and the combined field $\mathbf{E} + \mathbf{E}_H$ (see Figure 15.3). Using Equaation (15.3), we obtain

$$\cot \theta_H \equiv \frac{E}{E_H} = \frac{E}{v_d B} = \frac{m^*}{qB}\gamma_0. \tag{15.8}$$

15.2.4 Dynamic Coefficients

Let us now consider the *dynamic*, or *frequency-dependent*, case. A monochromatic IR radiation with angular frequency ω and k-vector $\mathbf{k}$ carries an oscillating electric field represented by

$$\mathbf{E}(\mathbf{r}, t) = \mathbf{E}\, e^{i(\mathbf{k}\cdot\mathbf{r}-\omega t)}, \qquad \mathbf{E}\cdot\mathbf{k} = 0. \tag{15.9}$$

The E-field running in the **k**-direction generates a running-wave current in the same direction. The wavelength $\lambda = 2\pi/k$ is much greater than the electron wave packet size. Hence we may omit the k-dependence of the response. We assume the *generalized Ohm's law* in the form:

$$\text{Running-wave part of } \mathbf{j} = \boldsymbol{\sigma}(\omega) \cdot \mathbf{E}\, e^{-i\omega t}, \tag{15.10}$$

where $\boldsymbol{\sigma}(\omega)$ is a dynamic conductivity tensor, which is necessarily complex.

A solid's response to radiations such as light and X-ray is known to depend on frequency ω. We set up a *generalized Drude equation* of motion for the conduction electron subject to an impurity damping (γ) and the sinusoidally oscillating electric field:

$$\frac{dv_x}{dt} + \gamma(\omega) v_x = (m^*)^{-1} q E e^{-i\omega t}. \tag{15.11}$$

In the large ω limit Equation (15.11) is reduced to the original Drude equation without γ. In the static limit the time derivative (dv_x/dt) vanishes. After taking the angle average, we obtain

$$v_d = \langle v_x \rangle = m^{*-1} q E \gamma^{-1}, \tag{15.12}$$

which is in agreement with Equation (15.3) if we choose

$$\gamma(0) \equiv \gamma_0 = \frac{1}{\tau}. \tag{15.13}$$

Hence, we see that Equation (15.11) describes both limits correctly.

We look for an oscillatory steady-state solution ($v_x \propto e^{-i\omega t}$) of Equation (15.11) and compare the terms proportional to $e^{-i\omega t}$. After using Equation (15.10), we obtain (Problem 15.2.1)

$$\boxed{\sigma(\omega) = \frac{q^2 n}{m^*} \frac{1}{\gamma - i\omega}.} \tag{15.14}$$

Comparison between Equation (15.14) and Equation (15.5) yields the following *conversion rule* for the dynamic formula:

$$\boxed{\gamma_0 \rightarrow \gamma(\omega) - i\omega\,.} \tag{15.15}$$

The dynamic rate γ in general is complex. For the following developments of the theory we assume a real γ. Using Equation (15.14), we obtain

$$\sigma(\omega) = \frac{q^2 n}{m^*} \frac{1}{\gamma - i\omega} = \frac{q^2 n}{m^*} \frac{1}{\gamma^2 + \omega^2} (\gamma + i\omega). \tag{15.16}$$

Let us take the ratio

$$\boxed{\frac{\mathrm{Re}[\sigma(\omega, T)]}{\mathrm{Im}[\sigma(\omega, T)]} = \frac{\gamma(\omega, T)}{\omega},} \tag{15.17}$$

where Equation (15.16) was used. If the real and imaginary parts of the dynamic conductivity $\sigma(\omega, T)$ are measured, Equation (15.17) allows a *direct* determination of the (T- and ω-dependent) rate $\gamma(\omega, T)$. This is significant. In contrast, the static conductivity formula $\sigma = q^2 n (m^*)^{-1} \gamma_0^{-1}$ contains two unknowns (m^*, γ_0) and, hence cannot yield γ_0 by itself.

Next, we apply the rule (15.15) to formula (15.8) and obtain

$$\cot \theta_H \simeq \theta_H^{-1} = \frac{m^*}{qB} [\gamma_H(\omega, T) - i\omega]. \tag{15.18}$$

Hence we obtain

$$\mathrm{Re}[\theta_H^{-1}] = \frac{m^*}{qB} \gamma_H(\omega, T), \tag{15.19}$$

$$\mathrm{Im}[\theta_H^{-1}] = -\frac{m^*}{qB} \omega. \tag{15.20}$$

The ratio of the real and imaginary parts is then

$$\boxed{\frac{\mathrm{Re}[\theta_H^{-1}]}{\mathrm{Im}[\theta_H^{-1}]} = -\frac{\gamma_H(\omega, T)}{\omega},} \tag{15.21}$$

which gives the *dynamic Hall relxation rate* $\gamma_H(\omega, T)$ *directly*.

Problem 15.2.1. Verify Equation (15.14).

15.3 Discussion

The observed IR $\mathrm{Re}[\theta_H^{-1}]$ is negative; this means that the charge carrier is the "electron":

$$\mathrm{Re}[\theta_H] < 0 \quad \Rightarrow \quad q = -e. \tag{15.22}$$

The observed T dependence represented by Equation (15.1) can simply be explained based on formula (15.19). The T linear behavior of $-$IR $\mathrm{Re}[\theta_H^{-1}]$ simply means that

$$\gamma_H \propto T. \tag{15.23}$$

This T behavior arises from the phonon scattering, since the phonon population is proportional to the absolute temperature T in the measured range (20–230 K).

The observed ω-dependence represented by Equation (15.2) can be interpreted based on formula (15.20). The observed IR $\mathrm{Im}[\theta_H^{-1}]$ is positive in agreement with formula (15.20) with $q = -e$, confirming that the carrier is the "electron." Cerne et al. [1] measured the static Hall angle $\theta_H(\omega = 0, T) \equiv \theta_H(T)$, and compared the data with IR $\mathrm{Re}[\theta_H^{-1}]$. The triangle ▼ in Figure 15.4 (a) indicates the experimental point for static $\theta_H(T)$. Both data are on the single line shown. This means that

$$\gamma_H(\omega, T) = \gamma_{H,0}(T). \tag{15.24}$$

The dynamic Hall relaxation rate is equal to the static Hall rate up to the IR frequency, which is *most remarkable*.

Our theory allows us to obtain the *absolute value* of the rate $\gamma_H(\omega, T)$ based on Equation (15.17). In the measurements shown in Figure 15.3, the ratio $\mathrm{Re}[\theta_H^{-1}]/\mathrm{Im}[\theta_H^{-1}] = \gamma_H/\omega$ at 8 T varies linearly in T from 1/4 to 2/5 as the temperature is raised from 20 to 230 K. This T-linear behavior is caused by the change in the phonon scattering rate $\gamma_H(\omega, T)$ [see Equation (15.22)]. The ratio $-\mathrm{Re}[\theta_H^{-1}]/\mathrm{Im}[\theta_H^{-1}]$ at $B = 8$ T and 290 K changes inverse linearly with frequency ω. The slope of the solid line for $\mathrm{Im}[\theta_H^{-1}]$ versus the frequency expressed in the wave-number units (cm^{-1}) is 0.188, see Figure 15.3 (b). Translated against the frequency per second, the slope value is multiplied by the factor 3×10^{10} (light speed $c = 3\times 10^{10}$ cm/sec). Using Equation (15.21), we obtain

$$\begin{aligned} \gamma_H &= -\frac{\mathrm{Re}[\theta_H^{-1}]}{\mathrm{Im}[\theta_H^{-1}]}\omega \\ &= \frac{90}{0.188} \times 3 \times 10^{10} \ \mathrm{sec}^{-1} \\ &= 1.44 \times 10^{13} \ \mathrm{sec}^{-1} \quad (T = 290 \ \mathrm{K}). \end{aligned} \tag{15.25}$$

This value is obtained from all of the data for $\mathrm{IR}[\theta_H^{-1}]$ and $\mathrm{Im}[\theta_H^{-1}]$ in Figure 15.3, which indicates that the relaxation rate γ_H is 1.44×10^{13} sec^{-1} at 290 K.

The Hall coefficient R_H does not contain γ_0, and hence, $R_H(\omega)$ should remain to have the value $(qn)^{-1}$:

$$R_H(\omega) = \frac{1}{qn}. \tag{15.26}$$

This behavior is in agreement with the experimental data in Au and Cu [1].

The central simplifying feature of the kinetic theory of the static and dynamic Hall effect for a single-component system is the balanced force equation (see Figure 15.4):

$$E_H = v_d B, \tag{15.27}$$

where E_H and B are measurable. From this relation, the drift velocity $\mathbf{v}_d$ can be obtained *directly*. Using Equation (15.27), we obtain the cotangent of the Hall angle θ_H (see Figure 15.5) as

$$\cot\theta_H \equiv \frac{E}{E_H} = \frac{E}{v_d B}, \tag{15.28}$$

where E and E_H are measurable. In the dynamic case, the applied electric field E and the induced quantities (j, v_d, E_H) all oscillate in time with the same frequency ω.

Au (or Cu) is known to have "necks" at the Brillouin boundary [3]. Hence it is not a one-carrier metal, and our formulas are not strictly applicable. As far as $\theta_H(\omega, T)$ and $R_H(\omega, T)$ are concerned, our kinetic theory for the one-carrier model is in good agreement with the experiments.

The static rate γ_0 and the Hall rate γ_H are the same for the Drude model. The clearly different values for γ_0 and γ_H observed in Au and Cu are likely to be due to the carrier differences, "electrons" and field-dressed electrons, as discussed in Section 12.3 [1].

Our main theoretical results, Equations (15.14) and (15.15), are applicable not only to fermionic electrons but also to bosonic carriers such as Cooper pairs. This is important when dealing with the IR Hall effect in cuprates (high-temperature superconductors).

Appendix A

Electromagnetic Potentials

The electromagnetic force is one of the forces whose nature is most firmly understood. It is also one of the most important forces acting on electrically charged particles. The *Lorentz force* acting on a particle of charge q moving with velocity $\dot{\mathbf{r}}$ is given by

$$\mathbf{F} = q(\mathbf{E} + \dot{\mathbf{r}} \times \mathbf{B}). \tag{A.1}$$

The Lorentz force depends on the velocity, and hence, it is not a conservative force. However, it can be generated from a *velocity-dependent potential* U such that the correct equation of motion is obtained in the standard form

$$\frac{d}{dt}\left(\frac{\partial L}{\partial \dot{q}_j}\right) - \frac{\partial L}{\partial q_j} = 0, \qquad j = 1,\, 2,\, 3, \tag{A.2}$$

with the Lagrangian defined by $L \equiv T - U$. This will be shown here.

It is known in electromagnetic theory that the electric and magnetic fields ($\mathbf{E}$, $\mathbf{B}$) can be derived from *scalar* and *vector potential fields* (ϕ, $\mathbf{A}$) through the relations

$$\mathbf{E} = -\nabla\phi(\mathbf{r}, t) - \frac{\partial}{\partial t}\mathbf{A}(\mathbf{r}, t); \quad E_x = -\frac{\partial \phi}{\partial x} - \frac{\partial A_x}{\partial t}, \ldots, \tag{A.3}$$

$$\mathbf{B} = \nabla \times \mathbf{A}; \quad B_y = \frac{\partial A_x}{\partial z} - \frac{\partial A_z}{\partial x}, \ldots. \tag{A.4}$$

Let us take the function

$$\begin{aligned} U &\equiv q\phi(\mathbf{r}, t) - q\dot{\mathbf{r}} \cdot \mathbf{A}(\mathbf{r}, t) \\ &\equiv q\phi - q(\dot{x}A_x + \dot{y}A_y + \dot{z}A_z) = U(x, y, z, \dot{x}, \dot{y}, \dot{z}). \end{aligned} \tag{A.5}$$

Differentiating it with respect to x and $\dot{x}$, we obtain

$$-\frac{\partial U}{\partial x} = -q\frac{\partial \phi}{\partial x} + q\left(\dot{x}\frac{\partial A_x}{\partial x} + \dot{y}\frac{\partial A_y}{\partial x} + \dot{z}\frac{\partial A_z}{\partial x}\right),$$

$$\frac{\partial U}{\partial \dot{x}} = -qA_x. \tag{A.6}$$

Further we obtain

$$\frac{d}{dt}\left(\frac{\partial U}{\partial \dot{x}}\right) = -q\frac{d}{dt}A_x(x, y, z, t)$$
$$= -q\left(\frac{\partial A_x}{\partial t} + \frac{\partial A_x}{\partial x}\dot{x} + \frac{\partial A_x}{\partial y}\dot{y} + \frac{\partial A_x}{\partial z}\dot{z}\right). \tag{A.7}$$

Using Equations (A.6) and (A.7), we obtain

$$-\frac{\partial U}{\partial x} + \frac{d}{dt}\left(\frac{\partial U}{\partial \dot{x}}\right) = q\left(-\frac{\partial \phi}{\partial x} - \frac{\partial A_x}{\partial t}\right)$$

$$+ q\left[\dot{y}\left(\frac{\partial A_y}{\partial x} - \frac{\partial A_x}{\partial y}\right) + \dot{z}\left(\frac{\partial A_z}{\partial x} - \frac{\partial A_x}{\partial z}\right)\right]$$

$$= qE_x + q\left(\dot{y}B_z - \dot{z}B_y\right)$$

$$= q\left(\mathbf{E} + \dot{\mathbf{r}} \times \mathbf{B}\right)_x. \tag{A.8}$$

The quantity in the last member is just equal to the x-component of the Lorentz force in Equation (A.1).

Let us now define the *generalized Lagrangian function* L by

$$L \equiv T - U$$

$$= \frac{1}{2}m(\dot{x}^2 + \dot{y}^2 + \dot{z}^2) + q(\dot{x}A_x + \dot{y}A_y + \dot{z}A_z) - q\phi$$

$$= \frac{1}{2}m\dot{r}^2 + q\dot{\mathbf{r}} \cdot \mathbf{A} - q\phi. \tag{A.9}$$

If we apply Lagrange's equations, Equation (A.2), of the standard form, we obtain, using Equation (A.8),

$$0 = \frac{d}{dt}(m\dot{x}) - \frac{d}{dt}\left(\frac{\partial U}{\partial \dot{x}}\right) + \frac{\partial U}{\partial x} = m\ddot{x} - q(\mathbf{E} + \dot{\mathbf{r}} \times \mathbf{B})_x, \qquad \text{(A.10)}$$

which is in agreement with Newton's equation of motion.

Let us now define the *canonical momentum* (p_x, p_y, p_z) by

$$p_x \equiv \frac{\partial L}{\partial \dot{x}}, \quad p_y \equiv \frac{\partial L}{\partial \dot{y}}, \quad p_z \equiv \frac{\partial L}{\partial \dot{z}}. \qquad \text{(A.11)}$$

Using the explicit form of the Lagrangian L in Equation (A.9), we obtain

$$p_x = m\dot{x} + qA_x, \quad p_y = m\dot{y} + qA_y, \quad p_z = m\dot{z} + qA_z$$

or in vector notation,

$$\boxed{\mathbf{p} = m\dot{\mathbf{r}} + a\mathbf{A}.} \qquad \text{(A.12)}$$

Notice that the canonical momentum $\mathbf{p}$ is distinct from the kinetic momentum constructed by the rule: mass times velocity.

Let us now introduce the Hamiltonian H in the standard manner. By expressing

$$H = \dot{x}p_x + \dot{y}p_y + \dot{z}p_z - L \equiv \dot{\mathbf{r}} \cdot \mathbf{p} - L \qquad \text{(A.13)}$$

in terms of the canonical variables (x, p_x, y, p_y, z, p_z), we obtain

$$H = \frac{1}{2m}\left[(p_x - qA_x)^2 + (p_y - qA_y)^2 + (p_z - qA_z)^2\right] + q\phi$$

or

$$\boxed{H \equiv \frac{1}{2m}|\mathbf{p} - q\mathbf{A}|^2 + q\phi.} \qquad \text{(A.14)}$$

Hamilton's equations of motion are

$$\dot{x} = \frac{\partial H}{\partial p_x}, \quad \dot{y} = \frac{\partial H}{\partial p_y}, \quad \ldots,$$

$$\dot{p_x} = -\frac{\partial H}{\partial x}, \quad \dot{p_y} = -\frac{\partial H}{\partial p_y}, \quad \ldots. \qquad \text{(A.15)}$$

Using the explicit form for the Hamiltonian H, we obtain, from the first set of Hamilton's equations of motion:

$$\dot{q}_j = \frac{\partial H}{\partial p_j}, \tag{A.16}$$

$$\dot{x} = \frac{\partial H}{\partial p_x} = \frac{1}{m}(p_x - qA_x), \quad \dot{y} = \frac{1}{m}(p_y - qA_y), \quad \ldots$$

or

$$\dot{\mathbf{r}} = \frac{1}{m}(\mathbf{p} - q\mathbf{A}), \tag{A.17}$$

which agrees with Equation (A.12). From the second set of Hamilton's equations,

$$\dot{p_j} = -\frac{\partial H}{\partial q_j}, \tag{A.18}$$

we obtain

$$\begin{aligned} \dot{p_x} &\equiv \frac{d}{dt}\{m\dot{x} + qA_x(x,y,z,t)\} \\ &= -\frac{\partial H}{\partial x} = \frac{q}{m}(\mathbf{p} - q\mathbf{A}) \cdot \frac{\partial \mathbf{A}}{\partial x} - q\frac{\partial \phi}{\partial x}. \end{aligned} \tag{A.19}$$

The reader may verify (Problem A.1) that this equation is equivalent to Newton's equation of motion:

$$m\ddot{x} = q(\mathbf{E} + \dot{\mathbf{r}} \times \mathbf{B})_x. \tag{A.20}$$

Problem A.1. Demonstrate the equivalence between Equations (A.19) and (A.20).

Appendix B

Statistical Weight for the Landau States

The statistical weight W for the Landau states in 3D and 2D will be calculated in this appendix.

B.1 The Three-Dimensional Case

Poisson's sum formula [1] is

$$\sum_{n=-\infty}^{\infty} f(2\pi n) = \frac{1}{2\pi} \sum_{m=-\infty}^{\infty} F(m) \equiv \frac{1}{2\pi} \sum_{m=-\infty}^{\infty} \int_{-\infty}^{\infty} d\tau f(\tau) e^{-im\tau}, \tag{B.1}$$

where F is the Fourier transform of f, and the sum $\sum_{n=-\infty}^{\infty} f(2\pi n + t)$, $0 \leq t < 2\pi$, is periodic with the period 1. The sum is by assumption uniformly convergent.

We write the sum in Equation (5.38) as

$$2\sum_{n=0}^{\infty} \sqrt{\epsilon - (2n+1)\pi} = (\epsilon - \pi)^{1/2} + \phi(\epsilon; 0), \tag{B.2}$$

$$\phi(\epsilon; x) \equiv \sum_{n=-\infty}^{\infty} (\epsilon - \pi - 2\pi|n + x|)^{1/2}. \tag{B.3}$$

Note that $\phi(\epsilon; x)$ is periodic in x with the period 1, and it can, therefore, be expanded in a Fourier series. After the Fourier series expansion, we set $x = 0$ and obtain Equation (B.2). By taking the real part (Re) of Equation (B.2) and using Equation (B.1) and Equation (5.38), we obtain

$$\left(A \frac{(\hbar\omega_c)^{3/2}}{\sqrt{2}\,\pi} \right)^{-1} W(E) = \frac{1}{\pi} \int_0^{\epsilon} d\tau (\epsilon - \tau)^{1/2} + \frac{2}{\pi} \sum_{m=1}^{\infty} (-1)^m \int_0^{\epsilon} d\tau (\epsilon - \tau)^{1/2} \cos m\tau, \tag{B.4}$$

where we assumed

$$\epsilon \equiv \frac{2\pi E}{\hbar\omega_0} \gg 1, \tag{B.5}$$

and neglected π against ϵ. The integral in the first term in Equation (B.4) yields $(2/3)\epsilon^{3/2}$, leading to W_0 in Equation (5.41). The integral in the second term can be written after integrating by parts, changing the variable ($m\epsilon - m\tau = t$), and using $\sin(A - B) = \sin A \cos B - \cos A \sin B$ as

$$\frac{1}{2m^{3/2}} \left(\sin m\epsilon \int_0^{m\epsilon} dt \frac{\cos t}{\sqrt{t}} - \cos m\epsilon \int_0^{m\epsilon} dt \frac{\sin t}{\sqrt{t}} \right). \tag{B.6}$$

We use asymptotic expansions for $m\epsilon = x \gg 1$:

$$\int_0^x dt \frac{\sin t}{\sqrt{t}} \sim \sqrt{\frac{\pi}{2}} - \frac{\cos x}{\sqrt{x}} - \cdots,$$
$$\int_0^x dt \frac{\cos t}{\sqrt{t}} \sim \sqrt{\frac{\pi}{2}} + \frac{\sin x}{\sqrt{x}} - \cdots. \tag{B.7}$$

The second terms in the expansions lead to W_L in Equation (5.42), where we used $\sin^2 A + \cos^2 A = 1$ and

$$\sum_{m=1}^{\infty} \frac{(-1)^{m-1}}{m^2} = \frac{\pi^2}{12}. \tag{B.8}$$

The first terms lead to the oscillatory term W_{osc} in Equation (5.43).

B.2 The Two-Dimensional Case

We write the sum in Equation (11.35) as

$$2\sum_{n=0}^{\infty}\Theta(\epsilon-(2n+1)\pi)=\Theta(\epsilon-\pi)+\psi(\epsilon;0), \tag{B.9}$$

$$\psi(\epsilon;x)\equiv\sum_{n=-\infty}^{\infty}\Theta(\epsilon-\pi-2\pi|n+x|). \tag{B.10}$$

Note that $\psi(\epsilon;x)$ is periodic in x and can therefore be expanded in a Fourier series. After the Fourier expansion, we set $x=0$ and obtain Equation (B.9). By taking the real part (Re) of Equation (B.9) and using Equation (B.1), we obtain

$$\mathrm{Re}\,[\text{Equation (B.9)}]=\frac{1}{\pi}\int_0^{\infty}d\tau\Theta(\epsilon-\tau)+\frac{2}{\pi}\sum_{\nu=1}^{\infty}(-1)^{\nu}\int_0^{\infty}d\tau\Theta(\epsilon-\tau)\cos\nu\tau, \tag{B.11}$$

where we assumed $\epsilon\equiv 2\pi E/\hbar\omega_c\gg 1$ and neglected π against ϵ. The integral in the first term in Equation (B.11) yields ϵ. The integral in the second term is

$$\int_0^{\infty}d\tau\Theta(\epsilon-\tau)\cos\nu\tau=\frac{1}{\nu}\sin\nu\epsilon. \tag{B.12}$$

Thus, we obtain

$$\mathrm{Re}\,[\text{Equation (B.9)}]=\frac{1}{\pi}\epsilon+\frac{2}{\pi}\sum_{\nu=1}^{\infty}\frac{(-1)^{\nu}}{\nu}\sin\nu\epsilon. \tag{B.13}$$

Using Equations (11.35) and (B.13), we obtain

$$\begin{aligned}W(E)&=W_0+W_{\text{osc}}\\&=C(\hbar\omega_c)\left(\frac{\epsilon}{\pi}\right)+C\hbar\omega_c\frac{2}{\pi}\sum_{\nu=1}^{\infty}\frac{(-1)^{\nu}}{\nu}\sin\left(\frac{2\pi\nu E}{\hbar\omega_c}\right).\end{aligned} \tag{B.14}$$

Appendix C

Derivation of Equation (11.19)

Let us consider an integral on the real axis

$$I(\alpha, R) = \int_{-R}^{R} dx \frac{e^{i\alpha(x+iy)}}{e^x + 1}, \qquad z = x + iy, \quad \alpha,\ R > 0. \tag{C.1}$$

We add an integral over a semicircle of the radius R in the upper z-plane to form an integral over a closed contour. We then take the limit as $R \to \infty$. The integral over the semicircle vanishes in this limit if $\alpha > 0$. The integral on the real axis, $I(\alpha, \infty)$, becomes the desired integral in Equation (11.19). The integral over the closed contour can be evaluated by using the residue theorem. Note that $(e^z + 1)^{-1}$ has simple poles at $z = i\pi,\ i3\pi,\ \ldots,\ i(2n-1)\pi,\ \ldots$. We may use the following formula valid for a simple pole at $z = z_j$:

$$\text{Res}\left(\frac{p(z)}{q(z)},\ z_j\right) = \frac{p(z_j)}{q(z_j)}, \tag{C.2}$$

where $p(z)$ is analytic at $z = z_j$, and the symbol Res means a residue [1]. We then obtain

$$\begin{aligned}
I(\alpha, \infty) &= 2\pi i \sum_{n=1}^{\infty} \text{Res}\left[\frac{e^{i\alpha z}}{e^{z} + 1},\ z_n = i(2n-1)\pi\right] \\
&= 2\pi i \sum_{n=1}^{\infty} \frac{e^{i\alpha[i(2n-1)\pi]}}{e^{i(2n-1)\pi}} \\
&= -2\pi i \frac{e^{-\alpha\pi}}{1 - e^{-2\alpha\pi}} \\
&= \frac{\pi}{i} \frac{1}{\sinh \alpha\pi}.
\end{aligned} \tag{C.3}$$

Appendix C

Derivation of Equation ([illegible])

References

Chapter 1

1. C. Kittel, *Introduction to Solid-State Physics*, 6th ed. (Wiley, New York, 1986).
2. P.A.M. Dirac, *Principles of Quantum Mechanics*, 4th ed. (Oxford University Press, London, 1958).
3. F. London, Nature **141**, 643 (1938); *Superfluids*, I and II (Dover, New York, 1964).
4. W. Pauli, Phys. Rev. **58**, 716 (1940).
5. P. Ehrenfest and J. R. Oppenheimer, Phys. Rev. **37**, 333 (1931); H. A. Bethe and R. F. Bacher, Rev. Mod. Phys. **8**, 193 (1936); S. Fujita and D. L. Morabito, Mod. Phys. Lett. B **12**, 753 (1998).

Chapter 2

1. A. Einstein, Ann. Physik **22**, 186 (1907).
2. P. Debye, Ann. Physik **39**, 789 (1912).
3. F.W. Sears and G.L. Salinger, *Thermodynamics, Kinetic Theory, and Statistical Mechanics*, 3rd ed. (Addison-Wesley, Reading, MA, 1975).

Chapter 3

1. A. Sommerfeld, Zeits. f. Physik **47**, 1 (1928).

Chapter 4

1. L.D. Landau, Zeits. f. Physik **64**, 629 (1930).

Chapter 5

1. W. Pauli, Zeits. f. Physik **41**, 81 (1927).
2. L.D. Landau, Zeits. f. Physik **64**, 629 (1930).

3. P.M. Morse and H. Feshbach, *Methods of Theoretical Physics* (McGraw-Hill, New York, 1953), pp. 466–467; R. Courant and D. Hilbert, *Methods of Mathematical Physics*, volume 1 (Interscience-Wiley, New York, 1953), pp. 76-77.

Chapter 7

1. F. Bloch, Zeits. f. Physik **52**, 555 (1928).
2. R.L. Kronig and W.G. Penney, Proc. Roy. Soc. (London) **130**, 499 (1931).
3. A. Haug, *Theoretical Solid-State Physics*, vol. 1 (Pergamon, Oxford, UK, 1972), pp. 64–69.
4. K. Shukla, *Kronig–Penney Models and their Applications to Solids*, PhD diss., State University of New York at Buffalo (1990).

Chapter 8

1. S. Fujita, S. Godoy and D. Nguyen, Found. Phys. **25**, 1209 (1995).

Chapter 9

1. W.A. Harrison, Phys. Rev. **118**, 1190 (1960).
2. W.A. Harrison, *Solid State Theory* (Dover, New York, 1979).
3. D. Schönberg and A.V. Gold, Physics of Metals-1, in *Electrons*, ed. J.M. Ziman (Cambridge University Press, Cambridge, UK, 1969), p. 112.
4. C. Kittel, *Introduction to Solid State Physics*, 6th ed. (Wiley, New York, 1986); N.W. Ashcroft and N.D. Mermin, *Solid State Physics* (Saunders, Philadelphia, 1976).

Chapter 10

1. A.H. Wilson, *Theory of Metals*, 2nd ed. (Cambridge University Press, London, 1953), pp. 50–51.
2. R. Becker, *Theorie der Elektrizität*, Vol. 2 (Teubner, Berlin, 1933), Sec. 56; see also Ref. 3, p. 194.
3. N.W. Ashcroft and N. D. Mermin, *Solid State Physics* (Saunders, Philadelphia, 1976), pp. 213–235.
4. G.H. Wannier, *Elements of Solid State Theory* (Cambridge University Press, London, 1960), pp. 190–194.
5. W.A. Harrison, *Solid State Theory* (Dover, New York, 1979), pp. 64–71.
6. C. Kittel, *Introduction to Solid State Physics*, 6th ed. (Wiley, New York, 1986), pp. 187–193.

7. A. Sommerfeld and H. Bethe, in *Handbuch der Physik* **24**, Part 2 (Springer, Berlin, 1933), pp. 333–622 (in particular, pp. 506–509); F. Seitz, *Modern Theory of Solids* (McGraw-Hill, New York, 1940), pp. 316–319.
8. P.A.M. Dirac, *Principles of Quantum Mechanics*, 4th ed. (Oxford University Press, London, 1958), pp. 121–125.
9. S. Fujita, S. Godoy, and D. Nguyen, Found. Phys. **25**, 1209 (1995).
10. L. P. Eisenhart, *Introduction to Differential Geometry* (Princeton University Press, Princeton, 1940).
11. T.W.B. Kibble, *Classical Mechanics* (McGraw-Hill, Maidenhead, England, 1966), pp. 166–171.
12. H. R. Ott et al., Phys. Rev. Lett. **52**, 1915 (1984); K. Hasselbach et al., Phys. Rev. Lett. **63**, 93 (1989).
13. S. Fujita and K. Ito, *Quantum Theory of Conducting Matter* 2. *Superconductivity and Quantum Hall Effect* (Springer-Verlag, New York), Chapters 19 and 20.

Chapter 11

1. W.J. de Haas and P.M. van Alphen, Leiden Comm. 208d, 212a (1930); Leiden Comm. 220d (1932).
2. L. Onsager, Phil. Mag. **43**, 1006 (1952).
3. M.R. Halse, Phil. Trans. Roy. Soc. A **265**, 507 (1969); D. Schönberg and A.V. Gold, Physics of Metals-1, in *Electrons* ed. J.M. Ziman (Cambridge University Press, Cambridge, UK, 1969), p. 112.
4. B.S. Deaver and W.M. Fairbank, Phys. Rev. Lett. **7**, 43 (1961).
5. R. Doll and M. Näbauer, Phys. Rev. Lett. **7**, 51 (1961).
6. L. Shubnikov and W.J. de Haas, Leiden Comm. 207a, 207c, 207d, 210a (1930).
7. H.L. Störmer et al., J. Vac. Sci. Thechnol. B, 1(2):423 (1983).
8. S. Godoy and S. Fujita, J. Eng. Sci. **29**, 1201 (1991).
9. W. Shockley, Phys. Rev. **90**, 491 (1953).
10. J. Wosnitza et al., Phys. Rev. Lett. **67**, 263 (1991).

Chapter 12

1. A.B. Pippard, *Magnetoresistance in Metals* (Cambridge University Press, Cambridge, UK, 1989), pp. 3–5.
2. J.R. Klauder and J.E. Kunzler, in *The Fermi Surface*, eds. Harrison and Webb (Wiley, New York, 1960).
3. L.W. Shubnikov and W.J. de Haas, Proc. Netherlands Royal Acad. Sci. **33**, 130 and 163 (1930).
4. F.E. Richard, Phys. Rev. B **8**, 2552 (1973).
5. M.A. Zudov et al., Phys. Rev. B **64** 201311 (2001).
6. R.G. Mani et al., Nature **420**, 646 (2002).
7. M.A. Zudov et al., Phys. Rev. Lett. **90**, 046807 (2003).

8. R.E. Prange and S.M. Girvin, eds., *The Quantum Hall Effect*, 2nd ed., (Springer-Verlag, New York, 1990); B.I. Halperin, P.A. Lee, and N. Read, Phys. Rev. B **47**, 7312 (1993).
9. R.G. Mani, Physica E **22**, 1 (2004).
10. R.B. Dingle, Proc. Roy. Soc. A **211**, 500 (1952).
11. D.C. Tsui, *4. DFG-Rundgespräch über den Quanten Hall Effekt*, (Schleching, Germany, 1989).
12. P. Ehrenfest and J.R. Oppenheimer, Phys. Rev. **37**, 333 (1931); H.A. Bethe and R.F. Bacher, Rev. Mod. Phys. **8**, 193 (1936); S. Fujita and D.L. Morabito, Mod. Phys. Lett. B **12**, 753 (1998).
13. S. Fujita et al., Phys. Rev. B **70**, 075304 (2004).
14. J.K. Jain, Phys. Rev. Lett. **63**, 199 (1989); Phys. Rev. B **40**, 8079 (1989); ibid. **41**, 7653 (1990); J.K. Jain, Surf. Sci. **263**, 65 (1992).
15. R.R. Du et al., Phys. Rev. Lett. **70**, 2944 (1993).
16. S. Fujita and K. Ito, *Quantum Theory of Conducting Matter* 2. *Superconductivity and Quantum Hall Effect* (Springer-Verlag, New York).

Chapter 13

1. G. Dresselhaus, A.F. Kip, and C. Kittel, Phys. Rev. **98**, 368 (1955).
2. A. Lempicki, D.R. Frankl, and V.A. Brophy, Phys. Rev. **107**, 1238 (1957).
3. E. Hückel, Zeits. f. Physik **70**, 204 (1931).
4. S. Fujita and S. Watanabe, Phys. Stat. Sol. (b) **159**, K69 (1990).
5. M.Y. Azbel and E.A. Kaner, J. Phys. Chem. Solids **6**, 113 (1958).
6. F.W. Spong and A.F. Kip, Phys. Rev. **137**, A431 (1965).
7. W. Shockley, Phys. Rev. **90**, 491 (1953).
8. M.S. Khaikin and R.T. Mina, Sov. Phys. JETP **15**, 24 (1962); ibid., **18**, 896 (1964).
9. Y. Onuki, H. Suematsu, and S. Tokura, J. Phys. Chem. Solids **38**, 419 (1977); ibid., 431 (1977).
10. S. Watanabe and S. Fujita, J. Phys. Chem. Solids **52**, 985 (1991); S. Fujita and S. Watanabe, Solid State Comm. **72**, 581 (1989).
11. M.P. Shaw, T.G. Eck, and D.A. Zych, Phys. Rev. **142**, 406 (1966).
12. J.J. Sabo, Phys. Rev. **118**, 1325 (1970).
13. D.F. Gibbons and L.M. Falicov, Phil. Mag. **8**, 177 (1963).

Chapter 14

1. P.L. Rossiter and J. Bass, *Metals and Alloys*, in *Encyclopedia of Applied Physics* **10**, VCH Publ. (1994), pp. 163–197.
2. S. Fujita, H-C. Ho, and Y. Okamura, Int. J. Mod. Phys. B **40**, 2254 (1989).
3. N.W. Ashcroft and N.D. Mermin, *Solid State Physics*, (Saunders, Philadelphia, 1976), pp. 256–258, 290–293.
4. D. Schönberg and A.V. Gold, Physics of Metals-1, in *Electrons*, ed. J.M. Ziman (Cambridge University Press, Cambridge, UK, 1969), p. 112.

Chapter 15

1. J. Cerne et al., Phys. Rev. B **61**, 8133 (2000).
2. S. Fujita, Y-G. Kim, and Y. Okamura, Mod. Phys. Lett. B **14**, 495 (2000).
3. N.W. Ashcroft and N.D. Mermin, *Solid State Physics* (Saunders, Philadelphia, 1976), pp. 225 and 240.

Appendix B

1. P.M. Morse and H. Feshbach, *Methods of Theoretical Physics* (McGraw-Hill, New York, 1953), pp. 466–467; R. Courant and D. Hilbert, *Methods of Mathematical Physics*, volume 1 (Interscience-Wiley, New York, 1953), pp. 76–77.

Appendix C

1. H.W. Wyld, *Mathematical Methods for Physics* (Benjamin, Reading, MA, 1976), pp. 450–451.

Bibliography

Solid State Physics

Ashcroft, N.W., and Mermin, N.D.: *Solid State Physics* (Saunders, Philadelphia, 1976).

Harrison, W.A.: *Solid State Theory* (Dover, New York, 1979).

Haug, A.: *Theoretical Solid State Physics*, I (Pergamon, Oxford, England, 1972).

Kittel, C.: *Introduction to Solid State Physics*, 6th ed. (Wiley, New York, 1986).

Seitz, F.: *Modern Theory of Solids* (McGraw-Hill, New York, 1940).

Sommerfeld, A., and Bethe, H.: *Handbuch der Physik* **24**, Part 2 (Springer, Berlin, 1933).

Wannier, G.H.: *Elements of Solid State Theory* (Cambridge University Press, London, 1960).

Wilson, A.H.: *Theory of Metals*, 2nd ed. (Cambridge University Press, London, 1953).

Background

Mechanics

Goldstein, H.: *Classical Mechanics* (Addison-Wesley, Reading, MA, 1950).

Kibble, T.W.B.: *Classical Mechanics* (McGraw-Hill, London, 1966).

Marion, J.B.: *Classical Dynamics* (Academic, New York, 1965).

Symon, K.R.: *Mechanics*, 3d ed. (Addison-Wesley, Reading, MA, 1971).

Quantum Mechanics

Alonso, M., and Finn, E.J.: *Fundamental University Physics, III Quantum And Statistical Physics* (Addison-Wesley, Reading, MA, 1989).

Dirac, P.A.M.: *Principles of Quantum Mechanics*, 4th ed. (Oxford University Press, London, 1958).

Gasiorowitz, S.: *Quantum Physics* (Wiley, New York, 1974).

Liboff, R.L.: *Introduction to Quantum Mechanics* (Addison-Wesley, Reading, MA, 1992).

McGervey, J.D.: *Modern Physics* (Academic Press, New York, 1971).

Pauling, L., and Wilson, E.B.: *Introduction to Quantum Mechanics* (McGraw-Hill, New York, 1935).

Powell, J.L., and Crasemann, B.: *Quantum Mechanics* (Addison-Wesley, Reading, MA, 1961).

Electricity and Magnetism

Griffiths, D.J.: *Introduction to Electrodynamics*, 2d ed. (Prentice-Hall, Englewood Cliffs, NJ, 1989).

Lorrain, P., and Corson, D.R.: *Electromagnetism* (Freeman, San Fransisco, 1978).

Wangsness, R.K.: *Electromagnetic Fields* (Wiley, New York, 1979).

Thermodynamics

Andrews, F.C.: *Thermodynamics: Principles and Applications* (Wiley, New York, 1971).

Bauman, R.P.: *Modern Thermodynamics with Statistical Mechanics* (Macmillan, New York, 1992).

Callen, H.B.: *Thermodynamics* (Wiley, New York, 1960).

Fermi, E.: *Thermodynamics* (Dover, New York, 1957).

Pippard, A.B.: *Thermodynamics: Applications* (Cambridge University Press, Cambridge, England, 1957).

Statistical Physics (undergraduate)

Baierlein, R.: *Thermal Physics* (Cambridge U.P., Cambridge, UK, 1999).

Carter, A.H.: *Classical and Statistical Thermodynamics* (Prentice-Hall, Upper Saddle River, NJ, 2001).

Fujita, S.: *Statistical and Thermal Physics*, I and II (Krieger, Malabar, FL, 1986).

Kittel, C., and Kroemer, H.: *Thermal Physics* (Freeman, San Francisco, CA, 1980).

Mandl, F.: *Statistical Physics* (Wiley, London, 1971).

Morse, P.M.: *Thermal Physics*, 2d ed. (Benjamin, New York, 1969).

Reif, F.: *Fundamentals of Statistical and Thermal Physics* (McGraw-Hill, New york, 1965).

Rosser, W.G.V.: *Introduction to Statistical Physics* (Horwood, Chichester, England, 1982).

Terletskii, Ya.P.: *Statistical Physics*, N. Froman, trans. (North-Holland, Amsterdam, 1971).

Zemansky, M.W.: *Heat and Thermodynamics*, 5th ed. (McGraw-Hill, New York, 1957).

Statistical Physics (graduate)

Davidson, N.: *Statistical Mechanics* (McGraw-Hill, New York, 1969).

Feynman, R.P.: *Statistical Mechanics* (Benjamin, New York, 1972).

Finkelstein, R.J.: *Thermodynamics and Statistical Physics* (Freeman, San Fransisco, CA, 1969).

Goodstein, D.L.: *States of Matter* (Prentice-Hall, Englewood Cliffs, NJ).

Heer, C.V.: *Statistical Mechanics, Kinetic Theory, and Stochastic Processes* (Academic Press, New York, 1972).

Huang, K.: *Statistical Mechanics*, 2d ed. (Wiley, New York, 1972).

Isihara, A.: *Statistical Physics* (Academic, New York, 1971).

Kestin, J., and Dorfman, J.R.: *Course in Statistical Thermodynamics* (Academic, New York, 1971).

Landau, L.D. and Lifshitz, E.M.: *Statistical Physics*, 3d ed., Part 1 (Pergamon, Oxford, England, 1980).

Lifshitz, E.M. and Pitaevskii, L.P.: *Statistical Physics*, Part 2 (Pergamon, Oxford, England, 1980).

McQuarrie, D.A.: *Statistical Mechanics* (Harper and Row, New York, 1976).

Pathria, R.K.: *Statistical Mechanics* (Pergamon, Oxford, England, 1972).

Robertson, H.S.: *Statistical Thermodynamics* (Prentice Hall, Englewood Cliffs, NJ).

Wannier, G.H.: *Statistical Physics* (Wiley, New York, 1966).

Index